The Galápag[os] Islands

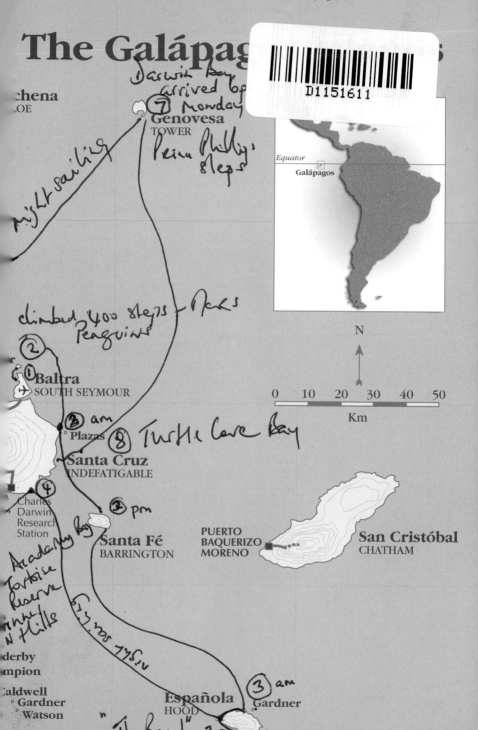

chena
LOE

Darwin Bay
arrived 6
⑦ Monday
Genovesa
TOWER

Prince Phillips
steps

night sailing

climbed 400 steps — Rocks
Penguins

②
① Baltra
✈ SOUTH SEYMOUR

⑥ am
Plazas

⑧ Turtle Cove Bay

Santa Cruz
INDEFATIGABLE

Charles
Darwin
Research
Station

⑤ pm

Santa Fé
BARRINGTON

PUERTO
BAQUERIZO
MORENO

San Cristóbal
CHATHAM

Academy Bay

Tortoise
Reserve

n Hills

erby
mpion

aldwell
  Gardner
  Watson

Española
HOOD
Gardner

③ am

"The Beach" 3 pm

Equator
Galápagos

N

| 0 | 10 | 20 | 30 | 40 | 50 |

Km

Sally Denham-Reid
10-23rd Feb '05

Birds, Mammals & Reptiles of the Galápagos Islands

Fernando Ortiz Q.
Galápagos Islands
2005
ferortizq@hotmail.com

" Caipirinha "

Andrew Vivens

Catherine Rowan

Juliet Vivens

Karen Thomson

Birds, Mammals & Reptiles of the Galápagos Islands

An identification guide

Andy Swash and Rob Still

With illustrations by Ian Lewington

WILDGuides

PICA PRESS

First published 2000 by Pica Press (an imprint of Helm
Information) and **WILD**Guides.

Pica Press   **WILD**Guides
The Banks   Parr House
Mountfield   63 Hatch Lane
Nr. Robertsbridge   Old Basing
East Sussex   Hampshire
TN32 5JY   RG24 7EB

www.wildguides.co.uk

ISBN 1-873403-82-8

Production and design by **WILD**Guides, Old Basing, Hampshire.
Printed by Midas Printing, Hong Kong.

CONTENTS

FOREWORD

Until the discovery of the Islands in 1535, the flora and fauna of Galápagos evolved in isolation, producing strange and marvellous island species, as Charles Darwin found when he visited the islands 300 years later. Nowadays many people can and do visit the islands to see the unique wildlife. One of the most remarkable things about Galápagos is that today's visitor can still encounter almost all the species that were present when Darwin and his predecessors came here. This is in striking contrast to the catastrophic chains of extinctions that have occurred in every other large oceanic archipelago following the arrival of man.

Modern visitors to Galápagos come to experience the wildlife and the lava landscapes and are invariably keen to identify the species they encounter. In many cases this experience fuels a desire to learn more about the relationships between the plants and animals and their environment. Such curiosity is amply rewarded when observing the thoroughly approachable, sometimes almost nonchalant Galápagos wildlife.

For many people, field guides provide the first insight into the diversity of wildlife and the intricacies and complexities of the natural world. This compact, well thought out and clearly presented book provides a comprehensive guide to the terrestrial vertebrates found in Galápagos. It is simple to use, the text is concise and the innovative use of digital technology has produced a remarkable collection of plates. An informative guide of this kind will help everyone from beginner to experienced naturalist.

With its ecological message – and the publisher's generous commitment to make a donation to the Charles Darwin Research Station for every copy sold - this book will also help conserve the Islands' wildlife. With over 15,000 people resident in the islands, daily flights, regular cargo boats from the mainland, and increasing local fishing pressure, the Islands no longer offer a safe, isolated haven to the vulnerable, endemic plants and animals. However, the Station, together with the Galápagos National Park Service, are working with skill and dedication to ensure that the encroachment of the modern world upon Galápagos is controlled and that the wildlife will continue to flourish. If you want to know more about our conservation, research and education work, or about the worldwide "Friends of Galápagos" network, then please visit our website at *www.galapagos.org*. I hope that this fine field guide will inspire you both to observe and to preserve the wonderful wildlife of Galápagos.

Dr Robert Bensted-Smith
Director
Charles Darwin Research Station

ACKNOWLEDGEMENTS

Many people have contributed to the production of this book and our sincere thanks are due to them all. It is our intention that everyone who has contributed to the book is named in this section, but if we have inadvertently missed anyone we can only apologise. Despite the contributions of others, we hold full responsibility for any errors which may have crept in, and for any omissions which may have been made.

With our limited knowledge of cetaceans, we would not have been able to write a definitive guide to the identification of the many species which have been recorded in Galápagos. We are, therefore, particularly indebted to Dylan Walker for his very significant contribution in writing this part of the book, and for collating the photographs which appear on the plates.

Condensing the plethora of information available on the birds of Galápagos, and defining their key identification features, proved to be particularly time-consuming and we are extremely grateful to Richard Kershaw for his tremendous assistance in helping to complete this task.

The plates are one of the key features of the book and are based largely on photographs taken by photographers from throughout the world. We would like to acknowledge the skill and patience of the following photographers who kindly allowed us to include their work in the book: Nigel Bean, Giff Beaton, Carla Benoldi, Kevin Carlson, Colin Carver, Roger Charlwood, John Cooper, David Cottridge, Graeme Cresswell, Yvonne Dean, Paul Doherty, Robert Dowler, Paul Ellis, Tom Ennis, Dagmar C. Fertl, John Geeson, Michael Gore, Bruce Hallett, Phil Hansbro, Brayton Holt, David Horwell, Barry Hughes, Boris Klusmeyer/Global Gecko Association, Gordon Langsbury, Tim Loseby, Mark Lucas, George McCallum/Whalephoto Berlin, Gary McCracken, Tony Marr, Lori Mazzuca, Jonathan Mitchell, Arthur Morris, Phil Myers/Animal Diversity Web, Dirk Neumann, Philip Newbold, Dick Newell, Howard Nicholls, Stephanie A. Norman, Dave Nye, Tony Palliser, Pietro Pecchioni, Mark Rauzon, Richard Revels, Chris Schenk, Hadoram Shirihai, Howard and Heidi Snell, Gill Swash, Alan Tate, Sam Taylor, David Tipling, Ralph Todd, Dylan Walker, Steve Young. The name of the photographer who took each photograph is shown in the photographic credits section of this book which starts on page 157. We would especially like to thank David Tipling of Windrush Photographs for his help in obtaining high quality photographs of some of the particularly difficult bird species, and to Tony Hutson for help with the bats.

Whilst we had originally intended to illustrate the book entirely with photographs, we were unable to source images of sufficiently high quality of all the species. There are very few illustrators who have the ability to paint birds accurately and make them appear lifelike: this was the reason for our wishing to use photographs in the first place. In our opinion, one of the best bird illustrators in the business is Ian Lewington and we were therefore delighted that he agreed to paint the plate of hirundines and Chimney Swift in flight (Plate 29), the measured drawings of the beaks of the Darwin's Finches (Plate 34) and the few 'missing' birds on the remaining plates. In this respect, we are also grateful to The Natural History Museum, Tring for allowing access to study the bird skins in their collection. Thanks are also due to Marc Dando for allowing us to use some of his artwork for the introduction to the cetaceans section and one of the plates.

Mavis and Len Still helped to type and format first drafts of much of the text. Robert Brooks, Mary Swash-Brooks and Richard Kershaw kindly proof-read the text at an early stage and Nigel Redman undertook the final edit of the book; all provided very helpful comments.

Thanks are also due to Mike Lindley and Sally-Ann Wilson for their help in sourcing reference material for the reptile section, to Amanda Hillier and Diana Langley of the Galápagos Conservation Trust, Howard and Heidi Snell at the Charles Darwin Research Station, Paul Coopmans, Hernan Vargas and Marco Altamirano for their help with various last-minute queries.

We are delighted that Rob Bensted-Smith, the Director of the CDRS, agreed to write the Foreword to the book, and are most grateful for his kind words.

Lastly, but by no means least, our wives, Gill and Penny, deserve special mention for their help, patience and understanding during the production of the book.

INTRODUCTION

About this book

This book aims to provide a handy yet comprehensive field guide to help visitors to Galápagos identify with confidence any bird, mammal or reptile they encounter.

The book focuses on the key identification features which separate each species from all others, highlighting where necessary the differences between males and females, and, in the case of birds, the different plumages. Summary information is provided on each species' status and habitat preferences and, where appropriate, behavioural characteristics and breeding seasons.

Each of the extant species of bird, mammal and reptile which has ever been recorded in Galápagos is covered, and all, bar a few of the lava lizards and leaf-toed geckos which are confined to single islands, are illustrated with photographs or paintings. The plates which form the bulk of the book are based on digitally manipulated images which have been resized to ensure that the species on each plate are in proportion.

We have chosen to be comprehensive in the coverage of the species in order to help visitors to the islands record all the species they see; it would be very easy to dismiss sightings of unusual species simply because they are not covered. We hope that by illustrating or describing them all (and in the case of birds their various plumages), visitors will be better able to contribute to the scientific record by submitting details of their sightings to the relevant authority. The Charles Darwin Research Station welcomes records of all scarce residents or migrants and vagrants. Details should be sent, including as full a description as possible, to the CDRS (see page 156 for contact details).

It is beyond the scope of this book to provide detailed information on the biology or ecology of each of the species, since such information is readily available in the many excellent books which have been published already on the wildlife of Galápagos. This book is an identification guide which we hope will stimulate further reading once the observer is confident of the identity of the birds, mammals and reptiles he or she encounters whilst exploring these fascinating islands.

How to use this book

The book is divided into seven main sections. The following two sections give a brief overview of the geography and climate (pages 10–11) and the habitats of Galápagos (pages 12–16), and emphasise the key factors that influence the distribution and breeding strategies of species in the archipelago.

The bulk of the book comprises stand-alone sections on the birds (pages 17–109), reptiles (pages 110–125) and mammals (pages 126–149). The first part of each of these sections provides an introduction to the 'types' of species that occur in Galápagos. This is followed by a series of plates, each with an accompanying page of text that summarises the status of each of the species shown and highlights the key identification features.

The final section of the book includes a checklist of the resident and regularly occurring species which summarises their habitat preferences and distribution (pages 150–154), a glossary of technical terms (page 155), suggestions for further reading (page 156), and a list of all the photographs and illustrations used in the book (pages 157–164).

THE GEOGRAPHY AND CLIMATE OF GALÁPAGOS

Geography

The Galápagos Islands are volcanic in origin and situated in the Pacific Ocean, straddling the equator, about 960km west of Ecuador in mainland South America and 1,100km from Costa Rica in Central America. The total land area of the archipelago is 7,882km² and the coastline extends to 1,336km.

The distance from the island of Darwin in the north-west to Española in the south-east is 430km; and it is 220km from Punta Cristóbal, the south-westernmost point of Isabela, to Genovesa in the north-east.

Nineteen of the islands have a surface area of over 1km² and 13 are over 10km². In addition, there are 42 islets with a surface area of less than 1km² and 26 emerging rocks. The largest island, Isabela, extends to 4,588km², representing over 70% of the land area. The highest point in Galápagos is Volcan Wolf on Isabela, which rises to 1,707m. The size and highest point of each of the main islands is shown in the following table. Five of the islands are inhabited and the resident population is approaching 17,000.

Island	Area (km²)	Altitude (m)	Inhabited
Isabela	4,588	1,707	Yes
Santa Cruz	986	864	Yes
Fernandina	642	1,494	No
Santiago	585	907	No
San Cristóbal	558	730	Yes
Floreana	173	640	Yes
Marchena	130	343	No
Española	60	206	No
Pinta	60	777	No
Baltra	27	100	Yes
Santa Fé	24	259	No
Pinzón	18	458	No
Genovesa	14	76	No
Rábida	4.9	367	No
Seymour	1.9	–	No
Wolf	1.3	253	No
Tortuga	1.2	186	No
Bartolomé	1.2	114	No
Darwin	1.1	168	No

Climate

The climate of Galápagos is unusually dry for the tropics, although there are considerable differences between the islands and much annual variation. However, there are two distinct seasons: the warm/wet season (December to May) and the Garúa (misty) or dry season

(May to December). During the warm/wet season the skies are usually clear and the temperature can exceed 30ºC, although it is humid and heavy rains occur. The Garúa season brings a subtropical climate, with the higher parts of the main islands often clothed in cloud for days on end. The direction of the prevailing south-easterly and north-easterly tradewinds means that the rainfall is concentrated on the southern and eastern slopes of the higher islands, the northern and western slopes being drier.

The average monthly temperatures (maximum and minimum), rainfall, and hours of sunshine are shown in the graphs opposite.

A number of ocean currents converge in Galápagos and these have a direct influence on the climate. The currents include the cold Peru Coastal (or Humboldt) and Peru Oceanic Currents from the south, driven by the south-easterly tradewinds, and the warm Niño (or Panamá) Flow from the

Average maximum and minimum monthly temperatures (°C)

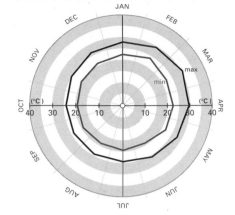

Average monthly rainfall (mm) and hours of sunshine

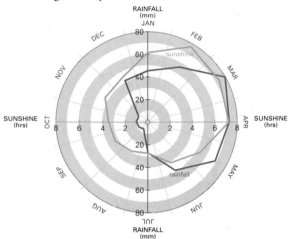

north, which is driven by the north-easterly tradewinds. These join the South Equatorial Current which passes the islands in a westerly direction. To the west of the archipelago this meets the cold, easterly-flowing Cromwell Current (or Equatorial Undercurrent) creating an upwelling that provides a rich source of food for many of the seabirds and marine mammals that occur in Galápagos. The timing of this upwelling is rather unpredictable and, as a consequence, many of the breeding species do so opportunistically.

In some years (on average one year in seven), the relative flow of warm currents is much greater than usual, resulting in an El Niño event. During such events, the surface temperature of the sea is much higher than usual and there is a very considerable increase in rainfall. This results in the rapid growth of vegetation and an increase in the populations of many of the landbirds. However, El Niño events also suppress the upwelling of cold waters, reducing the amount of food available for seabirds and marine mammals, the result of which is usually a significant decline in the populations of these species.

THE HABITATS OF GALÁPAGOS

The habitats of Galápagos can be divided into eight types, including the open sea and rocky islets. Six distinct vegetation zones are generally recognised which are broadly defined by altitude, reflecting the pattern of rainfall. Whilst these vegetation zones do differ from island to island, they are useful in describing the types of habitat in which the different species of birds and animals are found. The three vegetation zones which occur at the highest altitude are sometimes treated as one and termed the humid 'zone'. The term 'highlands' is used in this book to encompass all the vegetation zones from the transition zone upwards.

The open sea

The highly productive waters of Galápagos provide food for vast numbers of seabirds, as well as many species of cetacean (whales and dolphins). Some of the seabirds occur well away from land when they are feeding.

Seabird feeding frenzy – off Roca Redonda

Rocky islets

There are many sparsely vegetated rocky islets in Galápagos archipelago. The larger of these provide nesting sites for seabirds and some are home to Marine Iguanas and lava lizards.

Rocky islet – Roca Redonda

Seabird colony – Genovesa

The shore zone

The shore (or littoral) zone encompasses the range of habitats that occur close to the sea, including salt-water lagoons. These habitats include rocky shores, sea cliffs, sandy beaches and lagoons, as well as areas of mangroves, saltbush and Sea Purslane *Sesuvium* which are able to tolerate the saline conditions. The habitats of the shore zone support many seabirds, waterbirds and shorebirds as well some landbirds. The most obvious mammalian inhabitants are the sea lions, and the shore zone is also home to Marine Iguanas and turtles. In addition, lava lizards and leaf-toed geckos are found here, and snakes can occasionally be seen.

Rocky shore – Fernandina

Sandy beach and mangroves – Black Beach, Isabela

Mangroves – Turtle Cove, Santa Cruz

Lagoon bordered by saltbush – Rábida

The arid zone

The arid zone is the most diverse and extensive of all the vegetation zones in Galápagos. It occurs in the lowlands, extending inland from the shore zone, and is found on all of the main islands. The habitats that are typical of this zone include plant communities characterised by the presence of cacti and deciduous trees such as Palo Santo, Palo Verde and acacias. These habitats are home to virtually all of the species of landbird and also provide nesting sites for some of the seabirds, such as boobies and storm-petrels. Many of the reptiles, notably lava lizards and land iguanas, are at their most numerous in this zone, and rice rats also occur.

Prickly Pear cactus Opuntia *and Sea Purslane* Sesuvium *– South Plaza*

Palo Santo trees during the dry season – Isabela

The transition zone

The transition zone is characterised by a reduced number of the trees and shrubs which are typical of the arid zone and the presence of lichens and perennial herbs. The soils in this area are deeper than in the arid zone and the rainfall is higher. Most of the agricultural areas occur in the transition zone, with cattle grazing being the dominant form of land use. Many of the landbirds occur in this zone, as do Galápagos (or Giant) Tortoises.

Transition zone, agricultural area – Santa Cruz

Transition zone, disused agricultural area – Floreana

The humid 'zone'

The term humid 'zone' is often used to encompass the three distinct broad vegetation types that occur in the higher parts of the larger islands. These are the *Scalezia*, *Miconia* and pampa zones; a brief description of each follows.

The *Scalezia* zone
On the larger, higher islands the transition zone gives way to a moist, lush cloud-forest-type vegetation which consists of dense shrubs dominated by *Scalezia* trees. These trees, which are endemic to Galápagos, can reach a height of almost 20 metres and are festooned with mosses, liverworts, ferns, bromeliads and orchids. The *Scalezia* zone supports many species of landbird and provides breeding sites for the Dark-rumped Petrel.

The *Miconia* zone
The very wet, high altitude *Miconia* zone is found only on the southern slopes of Santa Cruz and San Cristóbal. It is treeless and occurs above the *Scalezia* zone, being characterised by dense stands of the endemic *Miconia* shrub, which grows to about two metres. Many species of grasses, liverworts and ferns are present in this zone which, due to its wetness, supports only a few species of landbirds.

The pampa zone
The pampa zone is the highest and wettest vegetation zone in Galápagos, usually occurring above 500 metres. Dominated by grasses, sedges and ferns, including the Galápagos Tree Fern that can grow to about three metres, the prevailing damp conditions mean that few species of birds, other than the ground-dwelling rails, are able to thrive.

Scalezia *zone – Santa Cruz*

Pampa zone – Isabela

THE BIRDS OF GALÁPAGOS

Introduction

In total, 152 bird species have been recorded in Galápagos. Only 61 of these are resident, although 28 are endemic to Galápagos (i.e. they are found nowhere else on earth) and a further 16 are represented by endemic subspecies. The other 91 species are either migrants that occur at least annually (25 species) or vagrants (66 species).

The species can be divided into eight 'types' which broadly reflect the habitats in which they occur. The following table lists these types and provides a summary of the number of species recorded and their status.

Type	Species Recorded	Status			Endemic Species	Endemic Subspecies
		Residents	Migrants	Vagrants		
Seabirds	47	19	4	24	5	8
Waterbirds	22	11	1	10	2	3
Shorebirds	34	2	16	16		1
Diurnal raptors	3	1	2		1	
Nightbirds	3	2		1		2
Larger landbirds	8	5	1	2	1	
Aerial feeders	6	1	1	4	1	
Smaller landbirds	29	20		9	18	2

It is a sad fact that seven of the resident species of Galápagos are threatened with extinction. Standard criteria for categorising the degree of threat that species face have been adopted by The International Union for Conservation of Nature and Natural Resources (the World Conservation Union) (IUCN). A summary of the definitions of these categories, and the species which are considered to fall into each, are shown in the following table.

Category	Definition	Galápagos species concerned
CRITICALLY ENDANGERED	Facing an extremely high risk of extinction in the wild in the immediate future.	Dark-rumped Petrel
ENDANGERED	Not critical but facing a very high risk of extinction in the wild in the immediate future.	Charles Mockingbird Mangrove Finch
VULNERABLE	Not endangered but facing a high risk of extinction in the wild in the medium-term future.	Galápagos Penguin Flightless Cormorant Galápagos Hawk Lava Gull

In addition, Elliot's Storm-petrel is considered to be DATA DEFICIENT (which means that inadequate information is available to assess the risk of extinction) and Waved Albatross, Wedge-rumped Storm-petrel, Galápagos Rail and Medium Tree Finch are considered NEAR-THREATENED (which means that, although they are at little risk of extinction in the medium-term future, they are still of conservation concern).

Bird topography

It is helpful to have some knowledge of the structure of a bird, and the names of the various parts, when attempting to identify a species and/or describe its features. For some species the colour or patterns of different groups of feathers can be particularly important in identification, and a knowledge of their arrangement is therefore also useful. The terms used in this book for the various parts of a bird, and the arrangement of the main feather tracts, are shown on the following photographs of Swallow-tailed Gull, Semipalmated Plover, Large Ground Finch and Chatham Mockingbird.

Swallow-tailed Gull

Hood
Leading edge
Secondaries
Greater coverts
Lesser coverts
Primaries
Wing-tip

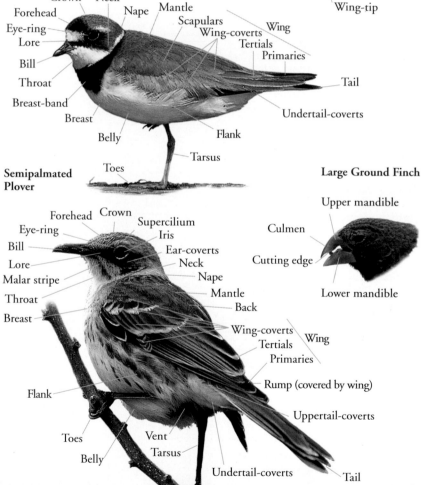

Crown Neck Mantle
Forehead Nape Scapulars
Eye-ring Wing-coverts Wing
Lore Tertials
Bill Primaries
Throat
Breast-band Tail
Breast Undertail-coverts
Belly Flank
Tarsus
Toes

Semipalmated Plover

Large Ground Finch

Upper mandible
Culmen
Cutting edge
Lower mandible

Forehead Crown
Eye-ring Supercilium
Bill Iris
Lore Ear-coverts
Malar stripe Neck
Throat Nape
Breast Mantle
Back
Wing-coverts Wing
Tertials
Primaries
Rump (covered by wing)
Flank Uppertail-coverts
Toes Vent
Belly Tarsus
Undertail-coverts Tail

Chatham Mockingbird

Moult and plumages

All birds replace their feathers – termed moult – at least once a year, the plumage often changing with successive moults. Some knowledge of the moult strategies of the different types of bird is therefore essential to the correct ageing of individuals, and in some cases is an invaluable aid to identification. Although moult strategies can be complex, most of the species within the groups of birds covered by this book follow a similar pattern. A summary of the sequences of plumages which occur is given in this section. In the New World, a different terminology is generally applied, and these alternatives are given in square brackets.

Juvenile [Juvenal] plumage
Juvenile plumage is the first plumage a bird attains once it has lost its down. In many species, juvenile plumage is very different from subsequent plumages. The juvenile feathers of many passerines have a rather 'loose' structure which often gives the bird a fluffy appearance. Most birds moult their juvenile feathers quite quickly after fledging and attain *first-winter* [*first basic*] plumage.

First-winter [First basic] plumage
First-winter plumage is the term used to describe the plumage most shorebirds, gulls and passerines attain after their first (*post-juvenile* [*first pre-basic*]) moult. During this moult it is usually only the body feathers which are replaced; the juvenile flight and tail feathers being retained (i.e. a *partial moult*). This can result in a distinctive plumage, notably in gulls.

First-summer [First alternate] plumage
Most species undergo a *pre-breeding* [*first pre-alternate*] *moult* during which at least some of the body feathers are replaced. This results in recognisable first-summer plumages with most shorebirds, gulls and passerines.

First adult [First definitive] plumage
Some birds, including pigeons and doves, undergo a *post-juvenile* [*first pre-basic*] *moult* during which all the feathers are replaced (i.e. a *complete moult*) and are subsequently inseparable from adults.

Subsequent immature plumages
A few species occurring in Galápagos take a number of years to attain adult plumage (e.g. certain gulls). These often have distinguishable immature plumages which chronologically are termed *second-winter* [*second basic*], *second-summer* [*second alternate*], and *third-winter* [*third basic*].

Adult [Definitive] plumage
Adult plumage is the final plumage a bird attains once it has reached maturity. Many species have just one adult plumage which is retained throughout the year, although others, notably the ducks, gulls and shorebirds, have different *breeding* [*alternate*] and *non-breeding* [*basic*] plumages. Adult birds adopt one of four moult strategies, depending upon the species, and this generally determines the differences between breeding and non-breeding plumages.

- *One complete post-breeding* [*pre-basic*] *moult per year* – e.g. most seabirds, diurnal (day-flying) raptors, nightbirds, pigeons and doves, and swifts.
- *One complete post-breeding* [*pre-basic*] *moult per year and a pre-breeding* [*pre-alternate*] *partial moult* – e.g. gulls, terns, skuas and jaegers, ducks, herons and egrets.
- *A partial post-breeding* [*pre-basic*] *moult and a complete pre-breeding* [*pre-alternate*] *moult* – some long-distance migrant passerines.
- *Two complete moults per year* – the only species recorded in Galápagos which are known to adopt this strategy are Franklin's Gull and Bobolink.

The types of bird

This part of the book provides an introduction to each of the eight 'types' of bird that occur in Galápagos and aims to aid initial identification. Each of these 'types' can be further subdivided into groups (which broadly equate to Families) and the following sections provide a description of each. The text for each group is cross-referenced to the relevant plate(s) which illustrate all the species ever to have been recorded. The species and subspecies within each group which are endemic to Galápagos are listed, the names of endemic species being shown in UPPER CASE text.

SEABIRDS

In total, 47 species of seabird have been recorded in Galápagos, 19 of which are resident. The seabirds therefore account for nearly one third of all the species ever recorded in the islands and about the same proportion of the resident species.

The seabirds can be conveniently divided into 12 groups, as shown in the following table. This shows the number of species recorded in each group and summarises their status. It also shows the number of endemic species and the number of other species which are represented by endemic subspecies. Species are treated as migrants if they occur annually, vagrants being those recorded less frequently.

Group (Plate No.)	Species Recorded	Status			Endemic Species	Endemic Subspecies
		Residents	Migrants	Vagrants		
Penguins (1)	1	1			1	
Cormorants (1)	1	1			1	
Pelicans (1)	1	1				1
Albatrosses (2)	4	1		3	1	
Shearwaters and petrels (2–4)	12	2	1	9		2
Storm-petrels (5–6)	8	3	1	4		2
Frigatebirds (7)	2	2				1
Boobies (8)	4	3		1		1
Skuas and jaegers (9)	2			2		
Gulls (9–10)	5	2	2	1	2	
Terns (11–12)	6	2		4		1
Tropicbirds (11)	1	1				
Total:	47	19	4	24	5	8

Galápagos Penguin (adult)

PENGUINS
Family: Spheniscidae
1 species recorded (resident)
(Plate 1)

Endemic species: GALÁPAGOS PENGUIN

Penguins are flightless, marine birds with thickset bodies; short, robust legs with webbed feet; and stout bills. The wings are compressed to form powerful, rigid flippers that are used for swimming underwater. When they are in the water, penguins can be difficult to see, as even when they are on the surface their bodies are held very low, often with only the head and neck visible. On land they appear rather clumsy, standing upright and walking with a waddling gait and occasional hops. The sexes are alike.

CORMORANTS
Family: Phalacrocoracidae
1 species recorded (resident)
(Plate 1)

Endemic species: FLIGHTLESS (or GALÁPAGOS) CORMORANT

Flightless Cormorant (adult)

Cormorants are largish waterbirds with elongated bodies; short, set-back legs with large, webbed feet; long necks; and long, hooked bills which are used to catch fish underwater. Cormorants can be difficult to see when they are in the water as even when they are on the surface they hold their body low, sometimes with only the head and neck showing. On land, they are rather ungainly, walking slowly and methodically. After returning to land from a period of feeding at sea they habitually hold their wings out to dry. The Flightless (or Galápagos) Cormorant is the largest and only flightless cormorant in the world. The sexes are alike and immature plumages resemble adult plumage.

PELICANS
Family: Pelecanidae
1 species recorded (resident)
(Plate 1)

Endemic subspecies: Brown Pelican

Brown Pelican (adult)

Pelicans are large, heavy waterbirds with short legs, webbed feet and exceptionally long bills. The bill has a large, distensible pouch which is used as a scoop-net to catch fish. In flight, pelicans hold their head and neck drawn back and have slow, rather ponderous wingbeats; they are, however, accomplished gliders. Pelicans are very ungainly on land, rarely walking far. The sexes are alike.

ALBATROSSES
Family: Diomedeidae
4 species recorded (1 resident; 3 vagrants)
(Plate 2)

Endemic species: WAVED ALBATROSS

Waved Albatross (adult)

Albatrosses are the largest of the seabirds, with extremely long, narrow wings which are held stiffly in flight. They have long, stout, hook-tipped bills with raised tubular nostrils, and shortish legs with large, webbed feet. Albatrosses are exceptional gliders, rarely flapping their wings other than in calm conditions. During such conditions they often sit on the sea for long periods. Albatrosses only come to land to breed and when they do they are somewhat ungainly, walking with a slow, rather waddling gait. The sexes are alike. The immature plumages of the Waved Albatross resemble the adult plumage (this is not the case with the other species recorded as vagrants to Galápagos).

SHEARWATERS and PETRELS
Family: Procellariidae
12 species recorded
(Plates 2–4)
(2 residents; 1 migrant; 9 vagrants)

Endemic subspecies: Audubon's Shearwater; Dark-rumped Petrel

Audubon's Shearwater (adult)

Although the species in the family Procellariidae range from the large giant petrels to the diminutive prions (which have both been recorded as vagrants to Galápagos), the species which are recorded regularly can be divided into two groups: shearwaters and petrels. All species in the family are long-winged, have short legs and webbed feet, and raised nostril

Dark-rumped Petrel (adult)

tubes at the base of the bill. Shearwaters are medium-sized, rather compact seabirds with long, slender bills and straight, stiffly-held wings in flight. Petrels are similar in size and shape to shearwaters but, in comparison, have short, deep bills and in flight tend to hold their wings slightly forward from the shoulder and sharply angled at the wrist or carpal joint. In all species recorded in Galápagos the sexes are alike and immature plumages resemble adult plumage.

Elliot's Storm-petrel (adult)

STORM-PETRELS
Family: Hydrobatidae
8 species recorded (Plates 5–6)
(3 residents; 1 migrant; 4 vagrants)

Endemic subspecies: Wedge-rumped Storm-petrel; Elliot's (or White-vented) Storm-petrel

Storm-petrels are the smallest of the seabirds with tube-noses (raised nostril-tubes at the base of the bill). They have short to medium-length wings in proportion to their size and usually fly close to the sea with a characteristic gliding flight action interspersed with rapid wingbeats. Storm-petrels have longish legs and webbed feet, and the species most likely to be seen in Galápagos are dark with white rumps. The sexes of all species recorded in Galápagos are alike, and immature plumages resemble adult plumage.

Magnificent Frigatebird (adult male)

FRIGATEBIRDS
Family: Fregatidae
2 species recorded (both resident) (Plate 7)

Endemic subspecies: Magnificent Frigatebird

Frigatebirds are large, dark seabirds with very long, pointed wings which are held forward from the shoulder in flight. They have deeply forked tails and short legs with webbed feet. Their bills are long and hook-tipped. Frigatebirds soar high over the sea and feed by harassing and kleptoparasitising other seabirds, particularly boobies, or by picking food from the surface of the sea. However, they never land on water. The male, female and immature plumages are distinctly different and breeding males have a bright red gular pouch which is inflated during display.

Nazca Booby (adult)

BOOBIES
Family: Sulidae
4 species recorded (3 residents; 1 vagrant) (Plate 8)

Endemic subspecies: Blue-footed Booby

Boobies are large, conspicuous seabirds with cigar-shaped bodies, long dagger-shaped bills and, in flight, long pointed wings and characteristic wedge-shaped tails. They have rather short legs but large webbed feet which, in the case of the Blue-footed and Red-footed Boobies, are used during courtship, the birds deliberately lifting their feet and showing them to their mates. The sexes are alike. Boobies feed at sea by plunge-diving from the air.

SKUAS and JAEGERS
Family: Stercorariidae
2 species recorded (both vagrants)   **(Plate 9)**

Skuas and jaegers are medium-sized to largish, rather heavy-bodied, brown or brown-and-white predatory seabirds with long, pointed, angled wings (broad in the case of *Catharacta* skuas) and hook-tipped bills. They are reminiscent of gulls but are readily distinguished in flight in all plumages by their dark wings with white flashes at the base of the primaries. The sexes are alike.

South Polar Skua (adult)

GULLS
Family: Laridae
5 species recorded   **(Plates 9–10)**
(2 residents; 2 migrants; 1 vagrant)

Endemic species: SWALLOW-TAILED GULL; LAVA GULL

Gulls are medium-sized to largish seabirds with long, pointed wings and longish, rather stout, hook-tipped bills, usually with a marked gonydeal angle. Their legs are longish and their feet webbed. The gulls recorded in Galápagos are predominantly grey and white in adult plumage, although the Lava Gull is wholly grey. The sexes are alike. Birds take

Swallow-tailed Gull (adult and juvenile

a number of years to attain adult plumage and immature plumages are variable, often making identification difficult. Gulls feed by picking food from the surface of the water or by scavenging, often along the shoreline.

TERNS
Family: Sternidae
6 species recorded (2 residents; 4 vagrants)   **(Plates 11–12)**
Endemic subspecies: Common Noddy

Terns are similar to gulls in many respects but are generally smaller, with narrower wings; thinner, straighter bills which lack the marked gonydeal angle; and shorter legs. Whilst terns are usually predominantly grey and white, the two species that breed in Galápagos are wholly dark brown (Common Noddy) and black and white (Sooty Tern). The sexes are alike. Terns feed by picking food from the surface of the water or by plunge-diving.

Common Noddy (adult)

TROPICBIRDS
Family: Phaethontidae
1 species recorded (resident)   **(Plate 11)**

Tropicbirds are predominantly white seabirds, resembling gulls and terns in size and structure. They are distinguished from these birds in adult plumage by their elongated central tail feathers and in all plumages by a black 'mask' through the eye. The bill is stout and slightly decurved and the legs are short with webbed feet. The sexes are alike. Tropicbirds feed at sea, flying low over the water and plunging to catch their food. They are often seen resting on the water with the tail raised.

Red-billed Tropicbird (adult)

In total, 21 species of waterbird have been recorded in Galápagos, ten of which are resident. The waterbirds can be conveniently divided into five groups, as shown in the following table. This shows the number of species recorded in each group and summarises their status. It also shows the number of endemic species and the number of other species which are represented by endemic subspecies.

Group (Plate No.)	Species Recorded	Status			Endemic Species	Endemic Subspecies
		Residents	Migrants	Vagrants		
Grebes (13)	1			1		
Ducks (13)	4	1		3		1
Rails and crakes (13,14)	6	3		3	1	
Flamingos (15)	1	1				
Herons and egrets (15-17)	10	6	1	3	1	2
Total:	22	11	1	10	2	3

Pied-billed Grebe (adult)

GREBES
Family: Podicepididae
1 species recorded (vagrant)
(Plate 13)

The only species of grebe recorded in Galápagos is the Pied-billed Grebe. This is a small, rather plump diving bird with a short neck; very short tail; straight, stout bill; and lobed toes. The sexes are alike.

White-cheeked Pintail (adults)

DUCKS
Family: Anatidae
4 species recorded (1 resident; 3 vagrants)
(Plate 13)
Endemic subspecies: White-cheeked Pintail

Ducks are medium-sized waterbirds with plump bodies, shortish necks with rounded heads, and medium-length, blunt-tipped and slightly flattened bills. Their legs are short and their feet webbed. Ducks are usually seen on the water or at rest on land. In flight, they are fast and direct, with rapid wingbeats, and the head and neck are held outstretched.

Galápagos Rail (adult)

RAILS and CRAKES
Family: Rallidae
6 species recorded (3 residents; 3 vagrants) (Plates 13–14)
Endemic species: GALÁPAGOS RAIL

Rails and crakes are small to medium-sized ground-dwelling birds with plump bodies; short tails, which are often held erect; shortish necks and bills; and long legs and toes (the feet are not webbed). They are generally rather furtive, mainly inhabiting marshy areas and are usually reluctant to fly. They are good swimmers. The sexes are alike in all species.

FLAMINGOS
Family: Phoenicopteridae
1 species recorded (resident)
(Plate 15)

Flamingos are large and unmistakable birds with extremely long legs and neck, and unique kinked bill. In adults the plumage is pink. The sexes are alike.

Greater Flamingo (adults)

HERONS and EGRETS   **Family: Ardeidae**
10 species recorded   **(Plates 15–17)**
(6 residents; 1 migrant; 3 vagrants)

Endemic species: LAVA (or GALÁPAGOS) HERON

Endemic subspecies: Great Blue Heron; Yellow-crowned Night-heron

Herons, egrets and night-herons are medium- to large-sized waterbirds with long necks and legs, unwebbed feet and long, straight, dagger-shaped bills. The smaller, white herons are usually referred to as egrets. The sexes are alike in all species.

Lava Heron (adult)

SHOREBIRDS

In total, 34 species of shorebird have been recorded in Galápagos, only two of which are resident. The shorebirds can be conveniently divided into four groups, as shown in the following table. This shows the number of species recorded in each group and summarises their status. None of the species is endemic and only one is represented by an endemic subspecies.

Group (Plate No.)	Species Recorded	Status			Endemic Species	Endemic Subspecies
		Residents	Migrants	Vagrants		
Oystercatchers (18)	1	1				1
Stilts (18)	1	1				
Plovers (19)	6		2	4		
Sandpipers, phalaropes and allies (18, 20–24)	26		15	11		
Total:	34	2	17	15		1

OYSTERCATCHER   **Family: Haematopodidae**
1 species recorded (resident)   **(Plate 18)**

Endemic subspecies: American Oystercatcher

Oystercatchers are large shorebirds with medium-length legs and long, orange bills. The only species recorded in Galápagos, American Oystercatcher, has a distinctive black and white plumage. The sexes are alike.

American Oystercatcher (adult)

Black-necked Stilt (adults)

STILTS
1 species recorded (resident)

Family: Recurvirostridae
(Plate 18)

Stilts are medium-sized shorebirds but are unmistakable due to their black and white plumage and exceptionally long legs. The sexes are similar.

*Semipalmated Plover
(adult, non-breeding)*

PLOVERS
6 species recorded (2 migrants; 4 vagrants)

Family: Charadriidae
(Plate 19)

Plovers are small to medium-sized, compact shorebirds with rather short necks; short, straight bills; and moderately long legs. They are readily told from other shorebirds by their behaviour, which involves periods of fast walking or running with sudden stops. The species recorded in Galápagos are from two distinct genera: the *Charadrius*

Black-bellied Plover (juvenile)

plovers are small and characterised by the presence of breast-bands, whereas the *Pluvialis* plovers are medium-sized and show black bellies in breeding plumage (although in non-breeding plumage they are relatively uniform grey or brown).

SANDPIPERS, PHALAROPES and ALLIES
26 species recorded (14 migrants; 12 vagrants)

Family: Scolopacidae
(Plates 18 and 20–24)

The other shorebirds recorded in Galápagos vary in size from the small sandpipers (often referred to as 'peeps') to the medium-sized Whimbrel. They also vary in structure from the short-legged to the long-legged and from the short-billed to the long-billed. The larger and strikingly plumaged species are generally quite easy to identify. However, the smaller ones can be difficult and some, particularly when they are in non-breeding plumage, are distinguishable only by careful observation of their overall structure, subtle plumage features, leg colour, bill shape and rump pattern.

Whimbrel (adult)

Wandering Tattler (adult)

Red-necked Phalarope (juvenile)

Just three species of diurnal (or day-flying) raptor have been recorded in Galápagos, one of which is an endemic resident. Each of the species represents a different group, as shown in the following table:

Group (Plate No.)	Species Recorded	Status			Endemic Species	Endemic Subspecies
		Residents	Migrants	Vagrants		
Osprey (25)	1		1			
Hawks (25)	1	1			1	
Falcons (25)	1		1			
Total:	3	1	2		1	

OSPREY — Family: Pandionidae
1 species recorded (migrant) (Plate 25)

The Osprey is a large, broad-winged bird of prey with a hooked bill, powerful legs and feet, and sharp, curved talons. They feed exclusively on fish and have reversible outer toes and spiny foot pads which help them grasp their slippery prey. They are readily identified by their white underparts and dark upperparts. The sexes are alike and immature plumages resemble adult plumage.

Osprey (juvenile)

HAWKS — Family: Accipitridae
1 species recorded (resident) (Plate 25)
Endemic species: GALÁPAGOS HAWK

Hawks are large, broad-winged birds of prey with hooked bills, a fleshy cere (the area at the base of the upper mandible), powerful legs and feet, and sharp, curved talons. The Galápagos Hawk is the only species to occur in Galápagos and is readily identified by its dark plumage. The sexes are alike although females are larger than males.

Galápagos Hawk (immature)

FALCONS — Family: Falconidae
1 species recorded (migrant) (Plate 25)

Falcons are medium-sized birds of prey with long, pointed wings which are often held angled at the wrist or carpal joint, hooked bills, a fleshy cere, relatively short but powerful legs and feet, and sharp, curved talons. The sexes are alike although females are larger than males.

Peregrine (adult)

NIGHTBIRDS

Just three species of nightbird have been recorded in Galápagos, two of which are resident with endemic subspecies. Each of the species represents a different group, as shown in the following table:

Group (Plate No.)	Species Recorded	Status			Endemic Species	Endemic Subspecies
		Residents	Migrants	Vagrants		
Barn owls (26)	1	1				1
Typical owls (26)	1	1				1
Nightjars (26)	1			1		
Total:	3	2		1		2

Barn Owl (adult)

BARN OWLS
1 species recorded (resident)

Family: Tytonidae
(Plate 26)

Endemic subspecies: Barn Owl

Barn Owls are medium-sized, mainly nocturnal, birds of prey with long, broad, rather rounded wings; hooked bills; relatively long, slender legs; and sharp, curved talons. They are distinguished from other owls by their white, heart-shaped facial discs. The sexes are alike and immature plumages resemble adult plumage.

TYPICAL OWLS
1 species recorded (resident)

Family: Strigidae
(Plate 26)

Endemic subspecies: Short-eared Owl

The typical owls are small to large-sized, mainly nocturnal, birds of prey with long, broad, rounded wings; hooked bills; relatively short, powerful legs; and sharp, curved talons. The only species to occur in Galápagos, the Short-eared Owl, is medium-sized and readily told from the Barn Owl by its mainly dark plumage, dark, circular facial disc and yellow eyes. The sexes are alike and immature plumages resemble adult plumage.

Short-eared Owl (adult)

Common Nighthawk (adult male)

NIGHTJARS
1 species recorded (vagrant)

Family: Caprimulgidae
(Plate 26)

Nightjars are medium-sized, mostly nocturnal, birds with long, pointed wings which are held forward from the shoulder and crooked in flight. They have very short, thin bills but large gapes which enable them to catch moths and other insects in flight. Nightjars have very short legs and rest during the day on the ground or on branches where they are difficult to see due to their cryptic plumage. The only species recorded in Galápagos is the Common Nighthawk which shows white wing-flashes in flight. The sexes are similar and immature plumages resemble adult plumage.

LARGER LANDBIRDS

Eight species of larger landbird have been recorded in Galápagos, five of which are resident including one endemic species. The larger landbirds can be conveniently divided into three groups, as shown in the following table:

Group (Plate No.)	Species Recorded	Status			Endemic Species	Endemic Subspecies
		Residents	Migrants	Vagrants		
Pigeons/doves (27)	3	2			1	
Kingfishers (27)	1		1			
Cuckoos (28)	4	3		1		
Total:	**8**	**5**	**1**	**2**	**1**	

Galápagos Dove (adult)

PIGEONS AND DOVES — Family: Columbidae

3 species recorded (2 residents, including **(Plate 27)** the introduced Feral Pigeon; 1 vagrant)

Endemic species: GALÁPAGOS DOVE

Pigeons and doves are small to medium-sized, rather plump birds with small heads which are bobbed back and forth when they are walking. They have relatively small, straight bills and short legs. In flight, pigeons and doves are fast and direct with rapid wingbeats. The sexes are alike and immature plumages resemble adult plumage.

Belted Kingfisher (adult female)

KINGFISHERS — Family: Alcedinidae

1 species recorded (migrant) **(Plate 27)**

Kingfishers are small to medium-sized, stocky birds with short legs and necks, large heads and very large, dagger-shaped bills which are used to catch fish. The only species recorded in Galápagos, the Belted Kingfisher, is large with unmistakable plumage.

CUCKOOS — Family: Cuculidae

4 species recorded (3 residents; 1 vagrant) **(Plate 28)**

Cuckoos are medium-sized, long-tailed, short-necked birds with long, pointed wings, resembling falcons in flight. They have short legs with two toes on each foot pointing forwards and two backwards. Most species of cuckoo have relatively short, stout bills, although the family also includes the black-plumaged anis which have large, ridged upper mandibles. Cuckoos are confined to the wetter, vegetated areas in Galápagos. The sexes are alike and immature plumages resemble adult plumage.

Dark-billed Cuckoo (adult)

AERIAL FEEDERS

Six species which feed exclusively on the wing (swifts, swallows and martins) have been recorded in Galápagos. Only one is resident, the endemic Galápagos Martin. The aerial feeders can conveniently be divided into two groups, as shown in the following table:

Group (Plate No.)	Species Recorded	Status			Endemic Species	Endemic Subspecies
		Residents	Migrants	Vagrants		
Swifts (29)	1			1		
Swallows and martins (29)	5	1	1	3	1	
Total:	6	1	1	4	1	

Chimney Swift

SWIFTS
Family: Apodidae
1 species recorded (vagrant)
(Plate 29)

Swifts are small to medium-sized, fast-flying aerial feeders with cigar-shaped bodies and long, stiffly-held, sickle-shaped wings which bend close to the body. They do not perch since their feet are adapted to clinging on rock faces etc., all four toes pointing forward. Swifts have very short, thin bills but large gapes with which they catch insects in flight.

Galápagos Martin (adult male)

SWALLOWS and MARTINS
Family: Hirundinidae
5 species recorded (1 resident; 1 migrant; 3 vagrants)
(Plate 29)

Endemic species: GALÁPAGOS MARTIN

Swallows and martins are small, aerial feeders with rather slender bodies and long, pointed wings which are held angled in flight, the wrist or carpal joint being further from the body than in swifts. They are extremely manœuverable and their short, pointed bills and wide gapes are well adapted for catching their invertebrate food in the air. Swallows and martins have very short legs but are able to perch.

SMALLER LANDBIRDS

In total, 29 species of smaller landbird have been recorded in Galápagos, 20 of which are resident. Eighteen of these resident species are endemic to Galápagos and the other two are represented by endemic subspecies. The smaller landbirds can be conveniently divided into ten groups, as shown in the following table:

Type (Plate No.)	Species Recorded	Status			Endemic Species	Endemic Subspecies
		Residents	Migrants	Vagrants		
Flycatchers (30)	3	2		1	1	1
New World warblers (31)	2	1		1		1
Vireos (31)	1			1		
Bananaquit (31)	1			1		
Waxwings (32)	1			1		
Grosbeaks (32)	2			2		
Tanagers (32)	1			1		
New World blackbirds (32)	1			1		
Mockingbirds (33)	4	4			4	
Darwin's finches (34–37)	13	13			13	
Total:	29	20		9	18	2

Vermilion Flycatcher (adult male)

TYRANT FLYCATCHERS
Family: Tyrannidae
3 species recorded (2 residents; 1 vagrant)   **(Plate 30)**

Endemic species: LARGE-BILLED (OR GALÁPAGOS) FLYCATCHER

Endemic subspecies: Vermilion Flycatcher

Flycatchers are small birds with large heads; short, broad-based, flat bills; and short legs. They have a rather erect posture and feed by making sallies from exposed perches to catch passing insects. The two species resident in Galápagos are readily identifiable.

Yellow Warbler (adult male)

NEW WORLD WARBLERS
Family: Parulidae
2 species recorded (1 resident; 1 vagrant)   **(Plate 31)**

Endemic subspecies: Yellow Warbler

New World warblers are small birds with short, thin, pointed bills. They inhabit vegetated areas and are generally very active, feeding by picking insects from the leaves or branches. The only resident species in Galápagos, the Yellow Warbler, is readily identified.

Red-eyed Vireo (adult)

VIREOS
Family: Vireonidae
1 species recorded (vagrant)   **(Plate 31)**

Vireos are small perching birds, reminiscent of warblers but rather more chunky with heavier bills. They are less active than warblers and the only species recorded in Galápagos, the Red-eyed Vireo, has a distinctive head pattern.

BANANAQUIT
Family: Coerebidae
1 species recorded (vagrant)   **(Plate 31)**

The Bananaquit is a small, warbler-like bird with a relatively long, thin, decurved bill. The plumage is distinctive with yellow underparts, dark upperparts and (in the adult) black and white head pattern.

Bananaquit (adult)

WAXWINGS
Family: Bombycillidae
1 species recorded (vagrant)   **(Plate 32)**

Waxwings are medium-sized, rather plump perching birds with prominent, sleek crests; short, broad-based bills; short legs; pointed wings; and relatively short tails. Their plumage is rather sombre although they have characteristic yellow-tipped tails and red waxy tips to their secondary feathers.

Cedar Waxwing (adult)

Rose-breasted Grosbeak (adult female)

GROSBEAKS
Family: Emberizidae
2 species recorded (both vagrants)   **(Plate 32)**

Grosbeaks are small to medium-sized, rather stocky perching birds with short, heavy bills; longish tails; and relatively short, broad wings. Although males of the two species recorded in Galápagos are readily identifiable in breeding plumage, females and immatures are less distinctive.

TANAGERS

Family: Emberizidae
Subfamily: Thraupinae

1 species recorded (vagrant)

(Plate 32)

Tanagers are medium-sized, fairly robust perching birds with toothed, conical bills. Adult males are generally brightly coloured and easily identified. Females and immatures of the only species recorded in Galápagos, the Summer Tanager, are rather dowdy.

Summer Tanager (adult female)

NEW WORLD BLACKBIRDS

Family: Aceridae

1 species recorded (vagrant)

(Plate 32)

The only species of New World blackbird recorded in Galápagos is the Bobolink. This species is medium-sized with a longish, rather ragged-tipped tail and short, conical bill. Males are readily identified in breeding plumage but females and juveniles are brown and streaked.

Bobolink (juvenile)

MOCKINGBIRDS

Family: Mimidae

4 species recorded (all resident)

(Plate 33)

Endemic species: GALÁPAGOS MOCKINGBIRD; CHARLES (OR FLOREANA) MOCKINGBIRD; HOOD MOCKINGBIRD; CHATHAM (OR SAN CRISTÓBAL) MOCKINGBIRD

Mockingbirds are medium-sized landbirds with long tails, longish legs and long, narrow, decurved bills. The plumage of the species in Galápagos is rather drab, with brownish upperparts and pale underparts.

Galápagos Mockingbird (adult)

DARWIN'S FINCHES

Family: Emberizidae
Subfamily: Emberizinae

13 species recorded (all resident)

(Plates 34–37)

Endemic species: SHARP-BEAKED GROUND FINCH; LARGE GROUND FINCH; MEDIUM GROUND FINCH; SMALL GROUND FINCH; LARGE TREE FINCH; MEDIUM TREE FINCH; SMALL TREE FINCH; VEGETARIAN FINCH; CACTUS FINCH; LARGE CACTUS FINCH; WOODPECKER FINCH; MANGROVE FINCH; WARBLER FINCH

Darwin's (or Galápagos) Finches are small landbirds with generally dull black, brown or olive, often streaky, plumage; short tails; and short, rounded wings. Their bills vary greatly in size and shape (a fact which was instrumental in inspiring Charles Darwin's thinking in relation to the theory of evolution - and hence the name given to this fascinating group of species). Darwin's Finches are found in all the habitats on Galápagos. Identification can be notoriously difficult due to the variation within each species and the occurrence of hybrids. For this reason, not all birds can be confidently identified to specific level and confusing individuals are probably best ignored!

Small Ground Finch (adult male)

Warbler Finch (adult)

The bird plates

Species order The order in which the species appear on the plates in this book does not follow precisely the taxonomic sequence adopted in most bird books. For ease of reference, the plates have instead been ordered so that all the birds of one 'type' (e.g. the seabirds) are together. Similarly, the species which are most likely to be confused appear, as far as possible, on the same plate. The text follows the sequence in which the species appear on the plate, from top to bottom. The species are numbered sequentially and the following annotations are used on the plates:

1 Adult in breeding plumage (species 1)

1n Adult in non-breeding plumage

1m Male   **1f** Female

1f-w First-winter   **1s-w** Second-winter

1i Immature   **1j** Juvenile

1c Chick

Distribution maps A distribution map accompanies the text for each of the resident species. To save space, three 'groups' of islands have been moved: the islands of Darwin and Wolf being shown in a small square at the top left of the map; Pinta, Marchena and Genovesa in the rectangle at the top right of the map; and San Cristóbal and Española in the rectangle to the bottom right. The true position of each of the islands is shown on the map which appears on the front endpaper of the book; this map also gives the names of each of the islands. A key to the colour codes used on the maps is shown on this 'dummy' map.

- Breeding range
- Non-breeding range
- Breeding colony
- Nesting area
- Regularly seen at sea
- Occasionally seen at sea
- Rarely seen at sea

English name The English names used in the book are those most commonly used in recent literature. Alternative names which are also used regularly are shown in brackets.

Spanish name The Spanish names shown are those which are most regularly used in Galápagos.

Endemic species and subspecies The names of species which are endemic to Galápagos are preceded by an **E** and those which are represented by endemic subspecies by an **e**.

Status The first part of the text for each species provides information on its status. This includes details of whether it is a resident, migrant or vagrant. For residents and migrants, a subjective assessment of abundance is given (and, where known, population estimates are shown). Endemic subspecies are named where appropriate and, for resident species, a summary of their breeding season(s) and habitat preferences is included.

Conservation status An indication of the conservation status of species which are threatened with extinction is given at the end of the first paragraph. The categories used are explained in the introduction to the birds on page 17, and are shown in RED TYPE.

Measurements The length of a bird is the measurement taken from the tip of the bill to the tip of the tail, with the bird lying flat on its back. For species which are regularly seen in flight a wingspan measurement is also included; this is the distance from wing-tip to wing-tip with the wings extended. Details of measurements are included in this book principally to give an indication of the relative size of each species.

Identification Details of the features which are key to the identification of the species are given, with brief descriptions of the different plumages where appropriate. Distinctions are made between male and female, breeding and non-breeding, and adult, immature and juvenile plumages where necessary. Where no such distinction is made, it should be assumed that the plumages are similar.

Voice The voices of birds are often difficult to describe since the perception of a call or song varies greatly from one person to another. However, where it is helpful to the identification of the species, details of relevant calls or songs are given.

Behaviour Particular behavioural characteristics which help in the identification of a species, such as the flight action of seabirds, are highlighted where appropriate.

33

PLATE 1

e **1** Brown Pelican *Pelecanus occidentalis* Pelicano Café

Length: 105–152 cm
Wingspan: 203–228 cm

Common resident; endemic subspecies *urinator*. Population estimated at a few thousand pairs; breeds throughout the year, nesting in small colonies in low bushes and mangroves, occasionally on the ground.

IDENTIFICATION: Unmistakable, due to very large size and long characteristic bill. ADULT: Sexes alike, although females generally slightly smaller than males. Plumage mainly grey-brown, with chestnut and white markings on head and neck. IMMATURE: Resembles adult but lacks head and neck markings and has pale underparts.

VOICE: Adults usually silent; young birds on the nest hiss and clap bill.

BEHAVIOUR: Often feeds by plunge-diving from the air, usually close inshore. Accomplished fliers, regularly seen soaring on thermals. Often fly to or from feeding grounds in orderly rows, with birds flapping and gliding almost in unison.

E **2** Flightless (or Galápagos) Cormorant *Phalacrocorax harrisi*

Cormorán no Volador

Length: 89–100 cm

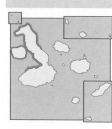

Uncommon, localised resident. Population estimated at *c.* 700–800 pairs in *c.* 100 small colonies, but declines dramatically after El Niño years. Nests on sheltered rocky shores, building bulky nest predominantly of seaweed just above the high-water mark. Breeds throughout the year, although most eggs are laid from March to September. Conservation Status: VULNERABLE.

IDENTIFICATION: Unmistakable; the only cormorant in Galápagos, with apparently functionless 'tatty' wings. ADULT: Sexes alike, although males noticeably larger than females; plumage dark blackish-brown above, paler brown below; eyes turquoise. IMMATURE: Resembles adult but has glossy-black plumage and dull brown eyes.

VOICE: Adults give a low growl; young birds on the nest give a plaintive "*wee-wee…*".

BEHAVIOUR: Flightless and marine; feeds by diving, usually close inshore.

E **3** Galápagos Penguin *Spheniscus mendiculus* Pingüino de las Galápagos

Length: 48–52 cm

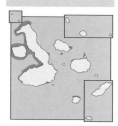

Uncommon, localised resident. Population fluctuates between a few thousand and a few hundred birds, declining dramatically after El Niño years. Nests in loose colonies in burrows or crevices close to the shore, breeding throughout the year depending on food availability. Conservation Status: VULNERABLE.

IDENTIFICATION: Unmistakable; the only penguin in Galápagos. ADULT: Sexes alike, although males are slightly larger than females. Upperparts, flippers and face black (brown when worn), with white line running through the eyes, down cheeks and across throat; underparts white with black line across breast and down flanks. IMMATURE: Resembles adult but greyer and lacks the head and chest pattern.

VOICE: A braying sound reminiscent of a donkey.

BEHAVIOUR: Flightless and pelagic, diving to feed.

PLATE 2

1 **Black-browed Albatross**   *Diomedea melanophris*   Albatros de Ceja Negr

Length: 83–93 cm
Wingspan: *c.* 240 cm

Vagrant; one record.

IDENTIFICATION: Size and shape as Waved Albatross, from which distinguished by diagnostic combination of black upperwing, back and tail, and white head and body with dark smudge through eye. ADULT: Underwing mainly dark with variable width white stripe; bill orange-yellow. IMMATURE: As adult but underwing and bill darker.

2 **Southern Giant Petrel**   *Macronectes giganteus*   Petrel Gigante Comú

Length: 86–99 cm
Wingspan: 185–205 cm

Vagrant; one record.

IDENTIFICATION: Large albatross-sized petrel, told from Wave Albatross by uniform grey-brown plumage with whitish face and breast (less obvious in juveniles, and silvery-grey underwing. At close range, prominent 'tube-nose' and yellow eye diagnostic Has a distinctive hump-backed appearance in flight.

3 **Wandering Albatross**   *Diomedea exulans*   Albatros Paseado

4 **Royal Albatross**   *Diomedea epomophora*   Albatros Rea

Length: 107–122 cm/
107–135 cm

Vagrant; two records, not specifically identified. Conservation Status o Wandering Albatross: VULNERABLE.

Wingspan: 254–351 cm/
305–351 cm

IDENTIFICATION: Extremely large albatrosses. The two species are separated with difficulty (see e.g. Harrison 1983). Wandering follows ships; Royal does not. ADULTS AND SUBADULTS (BOTH SPECIES) AND JUVENILE ROYAL: Distinguished from other albatrosses by diagnostic combination of wholly white head, body and back, and black upperwing with at least some white markings in centre or along leading edge. Underwing white with narrow black trailing-edge. At close range dark cutting edge to mandibles is diagnostic of Royal. JUVENILE WANDERING: Unique combination of brown body and upperwings, and white face.

5 **Black-footed Albatross**   *Diomedea nigripes*   Albatros de Piés Negros

Length: 68–74 cm
Wingspan: 193–213 cm

Vagrant; one confirmed record in the waters between Galápagos and the mainland; possible record from Española in 1897.

IDENTIFICATION: A small albatross, distinguished from Waved Albatross by diagnostic wholly dark brown plumage with white areas at base of bill, on uppertail-coverts (adults only) and often on undertail-coverts.

E **6** **Waved Albatross**   *Diomedea irrorata*   Albatros

Length: 85–93 cm
Wingspan: 230–240 cm

Localised resident, breeding only on Española. Total population estimated at *c.* 50,000–70,000 birds, including *c.* 12,000 breeding pairs. Breeds from March to January, nesting on the ground. Conservation Status: NEAR-THREATENED.

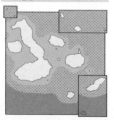

IDENTIFICATION: Unmistakable; due to large size and very long wings; the largest breeding bird in Galápagos. ADULT: Sexes alike, although males are generally slightly larger than females. Plumage mainly grey-brown with pale head and neck, and back of neck tinged yellow; underwings white with dark trailing-edge. Bill large and yellow. IMMATURE: Resembles adult but head whiter and bill dull.

VOICE: Silent at sea but gives a long, loud "*whoo–oo*" during display.

BEHAVIOUR: Pelagic. Very accomplished fliers, gliding for prolonged periods without flapping, particularly in windy conditions. Feeds from the surface, usually far out to sea. Performs bill-snapping and bill-rattling courtship display at breeding colony, mainly towards the end of the breeding season.

36

PLATE 3

1 **Parkinson's Black Petrel**   *Procellaria parkinsoni*   Petrel de Parkinso

Length: 46 cm
Wingspan: 115 cm

Vagrant; 3 specimens collected during an expedition in 1905 are the on
records. Conservation Status: VULNERABLE

IDENTIFICATION: A large, heavily-built petrel with buoyant flight on slightly bowe
wings. Plumage entirely blackish-brown except for pale shafts to primaries on underwing
Combination of pale bill with dark tip and black feet is diagnostic.

2 **Pink-footed Shearwater**   *Puffinus creatopus*   Pardela Negruzc

Length: 48 cm
Wingspan: 109 cm

Vagrant; recorded at sea off Isabela and Pinzón from October to Januar
Conservation Status: VULNERABLE

IDENTIFICATION: A large, heavily-built shearwater which flies in a rather 'lumbering
manner, with straight wings usually held forward. Plumage variable but generally greyish
brown upperparts, pale underparts, and palish mottled underwings with ill-defined dar
trailing-edge and tips. Combination of flesh-pink bill with dark tip and pink feet is diagnostic

3 **Sooty Shearwater**   *Puffinus griseus*   Pufino Negr

Length: 40–51 cm
Wingspan: 94–109 cm

Regular visitor in small numbers.

IDENTIFICATION: A large, stockily-built shearwater whic
usually flies fast with rather stiff wings; often 'towers' in strong winds. Plumage wholly darl
except for diagnostic silvery underwings. Bill dark.

4 **Wedge-tailed Shearwater**   *Puffinus pacificus*   Pardela del Pacífic

Length: 38–46 cm
Wingspan: 97–105 cm

Vagrant; one record of a bird found predated by a Short-eared Owl on Plazas
Occurs commonly in waters to the north of Galápagos so possibly overlooked

IDENTIFICATION: A large, long-winged shearwater which flies in a characteristic buoyant,
drifting manner with wings held bowed and well forward. Tail long and pointed. Occurs in
two colour morphs, dark and pale. The pale morph is most likely to occur and has brown
upperparts and crown, white underparts and white underwings with well-defined dark trailing-
edge and wing-tip. Bill uniform, grey to pink.

e **5** **Audubon's Shearwater**   *Puffinus lherminieri*   Pufino

Length: 27–33 cm
Wingspan: 64–74 cm

Common resident; endemic subspecies *subalaris*. Population estimated at
c. 10,000 pairs in about *c.* 30 colonies; breeds throughout the year, nesting
in crevices or burrows which are visited during the day.

IDENTIFICATION: A small shearwater; the only species breeding
in Galápagos. Blackish upperparts and crown, white underparts and
throat. Underwings white with dark trailing-edge and wing-tip.

VOICE: Usually silent at sea, but a loud "*kee–kaa–cooo*" is given
near the breeding colony.

BEHAVIOUR: Pelagic. Flight direct, often just skimming the surface
of the sea, with stiff wings and rapid wingbeats interspersed with
short glides. Often forms large feeding flocks, sometimes close
inshore. Feeds by plunge-diving from or close to the surface.

PLATE 4

❶ Cape Petrel *Daption capense* Petrel de Cabo

Length: 38–40 cm
Wingspan: 81–91 cm

Vagrant; one record in Galápagos waters. Regularly seen in waters between the archipelago and the mainland.

IDENTIFICATION: A medium-sized petrel. Distinctive, although variable, pattern of dark brown and white blotching on upperwing; underparts white with some dark spotting, but head dark.

❷ Southern Fulmar *Fulmarus glacialoides* Petrel Plateado

Length: 46–50 cm
Wingspan: 114–120 cm

Vagrant; one record.

IDENTIFICATION: A medium-sized, rather stockily-built petrel with characteristic gliding flight on stiff but slightly bowed wings. Upperwing, back and tail pale blue-grey with contrasting dark trailing-edge to wing, and dark outer-wing with variable pale patch at base of inner primaries; underparts white.

❸ Mottled Petrel *Pterodroma inexpectata* Petrel Moteado

Length: 33–35 cm
Wingspan: 74–82 cm

Vagrant; recorded at sea off Galápagos.

IDENTIFICATION: A medium-sized, long-winged petrel rarely seen close to land. Characteristic flight action involves high, bounding arcs. Distinguished from similar Dark-rumped Petrel by black 'W' shape across the upperwings; grey head, neck and rump; broad black line across underwings; and dark grey belly contrasting with white undertail-coverts.

❹ Antarctic Prion *Pachyptila desolata* Petrel Ballena

Length: 25–30 cm
Wingspan: 56–66 cm

Vagrant; only record is of a bird found dead on Floreana.

IDENTIFICATION: A small, blue-grey petrel with distinctive black 'M' mark across upperwing and dark tip to tail. Underparts white with dark mottling on sides of breast forming a partial collar. Underside of tail white with dark tip and central line. At close range dark 'mask' through eye and white mark above eye visible.

❺ Gould's (or White-winged) Petrel *Pterodroma leucoptera* Petrel de Gould

Length: 30 cm
Wingspan: 70–71 cm

Vagrant; recorded at sea off Galápagos.

IDENTIFICATION: A medium-sized, long-winged petrel rarely seen close to land. Characteristic flight action, at times reminiscent of Audubon's Shearwater, with a series of rapid wingbeats followed by a low, stiff-winged glide, but has sudden turns of speed involving jinking and banking in steep arcs before returning to flap–glide flight action. Distinguished from similar Dark-rumped Petrel by black 'W' shape across the upperwings, sooty-black crown and hindneck forming a distinct hood, and dark-tipped grey tail.

❻ Dark-rumped Petrel *Pterodroma phaeopygia* Patapegada

Length: 43 cm
Wingspan: 91 cm

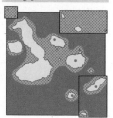

Uncommon resident; endemic subspecies *phaeopygia* often treated as a separate species: 'Galápagos Petrel'. Population estimated at 10,000–50,000 pairs in four colonies located in the highlands; breeds throughout the year, with different colonies nesting at different times; nests in burrows which are visited only at night. Conservation Status: CRITICALLY ENDANGERED

IDENTIFICATION: A large, long-winged petrel. Upperparts, crown and side of neck and breast uniform brownish-black; underparts and sides of rump white; conspicuous white forehead; underwing white with black line running along forewing and across centre of wing towards body. Small black mark in 'armpits' diagnostic.

VOICE: Silent at sea but calls at night near breeding colony "*kee–kee–kee— (c)ooo*".

BEHAVIOUR: Pelagic. Flight action characteristic: in calm weather involving a series of 3 or 4 flaps followed by a long glide on bowed and angled wings; in windy conditions glides in spectacular long arcs, high above the water, with wings bowed. Usually found well away from land during the day. Feeds from the surface whilst resting on the sea or by dipping in flight.

PLATE 5

e ① Wedge-rumped Storm-petrel *Oceanodroma tethys*

Golondrina de Tormenta de Galápág

Length: 18–20 cm
Wingspan: *c.* 50 cm

Common resident; endemic subspecies *tethys*. Population estimated *c.* 200,000 pairs in 3 colonies. Nests colonially in burrows or crevice breeding throughout the year but mainly during the cold season (April October). Conservation Status: NEAR-THREATENED.

IDENTIFICATION: A medium-sized, relatively long, narrow winged storm-petrel. Upperparts uniform dark brown with pal brown bar on upperwing and large, triangular-shaped, pure whit rump which extends almost to the tip of the tail and to the upp flanks and lateral undertail-coverts. Underparts dark, sometimes wit ill-defined pale centre to underwings.

BEHAVIOUR: Pelagic. Flight fast and rather forceful with deep wingbeats and involving muc banking and twisting, often high over the waves. Typically flies with wings held bowed an angled slightly forwards. When feeding skips and bounds over the surface, sometimes patterin on the water in a manner reminiscent of Elliot's Storm-petrel. Unique among storm-petrels i visiting the breeding grounds by day and feeding at night. Occasionally follows ships

② Madeiran (or Band-rumped) Storm-petrel *Oceanodroma castro*

Golondrina de Madeir

Length: 19–21 cm
Wingspan: 44–46 cm

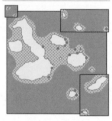

Uncommon resident. Population estimated at *c.* 15,000 pairs in *c.* 15 colonie Nests in burrows or crevices, breeding throughout the year, with two differen populations nesting in the same areas at different times.

IDENTIFICATION: A medium-sized storm-petrel with relativel broad, blunt-tipped wings; moderately long, slightly forked tail; an a rather 'bull-necked' appearance. The legs do not protrude beyon tail in flight. Upperparts uniform dark brown with paler brown ba on upperwing. Narrow, but prominent, 'U'-shaped white rum extends to the lateral undertail-coverts; at close range may show dark feathers at rear of white rump. Underparts entirely dark.

VOICE: Silent at sea but calls at night near the breeding colony: a squeaky "*whikka–whikka…*" rather like rubbing a wet finger on a glass.

BEHAVIOUR: Pelagic, feeding well offshore and rarely seen close to land. Flight action rather buoyant with rapid, shallow wingbeats and low, shearing glides reminiscent of Audubon's Shearwater, with the wings held flat or slightly bowed. Progresses in a zigzag manner bu occasionally becomes erratic, banking and 'jinking' and doubling back. Does not follow ships but sometimes attracted to lights at night. Visits colonies nocturnally.

e ③ Elliot's (or White-vented) Storm-petrel *Oceanites gracilis*

Golondrina de Tormenta de Elliot

Length: 15–16 cm
Wingspan: 40 cm

Common resident; endemic subspecies *galapagoensis*. Population estimated at many thousands and although a nest has yet to be found in Galápagos, breeding is suspected to occur between April and October. Conservation Status: DATA DEFICIENT.

IDENTIFICATION: A small storm-petrel with relatively short, broad-based and rather rounded wings, and a square-ended tail with feet protruding beyond tail in flight. Upperparts uniform dark brown with paler brown bar on upperwings and a narrow but prominent, 'U'-shaped, pure white rump which extends to the rear flanks. Underparts dark with pale grey patch on centre of belly, and ill-defined pale bar on underwings.

BEHAVIOUR: Pelagic, although often feeds close to shore. Feeds in a distinctive manner, fluttering and pattering on the water with wings raised. Often follows ships.

PLATE 6

❶ White-bellied Storm-petrel *Fregetta grallaria* Bailarí

Length: 19–20 cm
Wingspan: 46 cm

Vagrant; few records, mainly at sea to the south of Galápagos from May to August.

IDENTIFICATION: A medium-sized storm-petrel with rather broad, rounded wings and long legs which project just beyond tail. Characteristic direct flight, very close to the water surface with legs dangling and body swinging from side to side, often plunging breast-first into the water and then springing clear. Upperparts dark brown with prominent pale bar on upperwing contrasting with black flight feathers. Prominent, broad 'U'-shaped white rump extending to white breast and belly. Head and upper breast dark brown. Underwing coverts white, contrasting with dark flight feathers and leading edge of wing. Occasionally follows ships.

❷ White-faced Storm-petrel *Pelagodroma marina* Golondrina Cariblanca

Length: 20–21 cm
Wingspan: 41–43 cm

Vagrant; three records.

IDENTIFICATION: A medium-sized storm-petrel with broad, rather rounded wings, and feet projecting well beyond tail in flight. Flight action rather erratic, involving much banking and weaving often with jerky and rhythmic wingbeats somewhat reminiscent of Spotted Sandpiper. When feeding, 'dances' along the water surface with body swinging from side to side, the long legs being lowered only at the last minute before touching the surface. Upperparts grey-brown with pale bar on upperwing contrasting with black flight feathers; rump pale grey contrasting with black tail. Underparts and underwing-coverts white, contrasting with dark flight feathers. Brown and white face pattern distinctive. Rarely follows ships.

❸ Leach's Storm-petrel *Oceanodroma leucorhoa* Golondrina de Mar

Length: 19–22 cm
Wingspan: 45–48 cm

Vagrant; few records but possibly overlooked due to similarity with other species.

IDENTIFICATION: A largish storm-petrel with relatively long narrow wings which are usually held bowed and slightly forward. Flight buoyant and rather bounding with deep, tern-like wingbeats and low, shearing glides. Tail longish and forked, although fork not always visible. Upperparts dark brown with prominent pale brown bar on upperwing contrasting with dark flight feathers. Narrow, 'V'-shaped white rump with dark line up the centre (can be difficult to see). (Beware that dark-rumped birds may also occur and could be confused with Markham's or Black Storm-petrels.) Underparts wholly dark.

❹ Black Storm-petrel *Oceanodroma melania* Golondrina de Tormenta Negra

Length: 23 cm
Wingspan: 46–51 cm

Vagrant; two records but possibly overlooked due to confusion particularly with Markham's Storm-petrel. A warm-water species and therefore most likely to occur in El Niño years.

IDENTIFICATION: A largish, 'dark-rumped' storm-petrel with relatively long wings. Flight buoyant and rather direct, with deep wingbeats reminiscent of Black Tern (the wings being raised and lowered to *c.*60° above and below the horizontal) and occasional glides. Tail noticeably forked (although less deeply than Markham's). Upperparts dark brownish-black with prominent pale bar on upperwing not reaching the leading edge. Confusion likely only with Markham's and dark-rumped Leach's Storm-petrels, from which best separated on flight action. Occasionally follows ships.

❺ Markham's Storm-petrel *Oceanodroma markhami* Golondrina de Markham

Length: 23 cm
Wingspan: *c.* 50 cm

Vagrant; occasional records.

IDENTIFICATION: A largish, 'dark-rumped' storm-petrel with relatively long wings. Flight buoyant, with deliberate, shallow wingbeats (rarely more than 30° above and below the horizontal) and regular glides. Tail deeply and noticeably forked. Upperparts dark brown with prominent pale bar on upperwing reaching the leading edge and contrasting with dark flight feathers. Confusion likely only with Black and dark-rumped Leach's Storm-petrels, from which best separated on flight action. Does not normally follow ships.

PLATE 7

❶ Magnificent Frigatebird *Fregata magnificens*

Fragata Re

Length: 89–114 cm
Wingspan: 217–214 cm

Resident, endemic subspecies *magnificens*. Population estimated at *c.* 1,0(
pairs in 12 colonies; breeds throughout the year.

IDENTIFICATION: A large, dark, long-winged seabird with rakis
flight and long, deeply forked tail. Slightly larger than very simil
Great Frigatebird which is the only likely confusion species. ADU
MALE: Wholly black (apart from red gular sac); purplish sheen
mantle feathers, and black or brown legs and feet diagnostic. I
flight very difficult to distinguish from Great Frigatebird, althoug
usually lacks the pale bar across the upperwing which is typical
Great. ADULT FEMALE: Similar to male but breast white. Blue ey
ring diagnostic. In flight very similar to Great Frigatebird but whi
breast and black throat and thin white lines on axillaries are diagnostic. JUVENILE: Resembl
female but head as well as breast white.

Note that since birds do not reach maturity until their fourth year, gradually developir
adult plumage during this period, a series of intermediate plumages occur, often makir
specific identification difficult. Immature birds do, however, often show some thin whi
lines on the axillaries.

VOICE: Silent at sea, but male makes a drawn-out "*oo–oo–oo–…*" sound when displaying
the breeding colony.

BEHAVIOUR: Generally feeds close inshore, mainly by kleptoparasitising passing seabird
particularly Blue-footed Boobies. When displaying, males inflate their bright red gular sa
like a balloon and call to attract females, at the same time vibrating their outstretched wing

❷ Great Frigatebird *Fregata minor*

Fragata Comú

Length: 86–100 cm
Wingspan: 203–230 cm

Resident. Population estimated at a few thousand pairs in 12 colonies; breed
throughout the year.

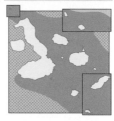

IDENTIFICATION: A large, dark, long-winged seabird with rakis
flight and long, deeply forked tail. Slightly smaller than very simila
Magnificent Frigatebird which is the only likely confusion species
ADULT MALE: Wholly black (apart from red gular sac); green sheer
to mantle feathers, and red or reddish-brown legs and feet diagnostic
In flight very difficult to distinguish from Magnificent, althoug
usually shows a pale bar across upperwing. ADULT FEMALE: Simila
to male but breast and throat white. Red or pink eye-ring diagnostic
In flight very similar to Magnificent Frigatebird but throat white
and does not show white markings on axillaries. JUVENILE: Resembles
female but head and breast usually washed pale orange or tawny is diagnostic. However
white-headed juveniles sometimes occur and are inseparable from juvenile Magnificent.

Note that since birds do not reach maturity until their fourth year, gradually developing
adult plumage during this period, a series of intermediate plumages occur, making separation
from Magnificent Frigatebird all the more difficult.

VOICE: Silent at sea but male gives a continuous rattling call when displaying at the breeding
colony.

BEHAVIOUR: A rather more pelagic species than Magnificent Frigatebird, generally only
being seen near land in the vicinity of the breeding colonies (which tend to be on the outer
islands of Galápagos). Feeds mainly by picking food from the surface although, like
Magnificent, regularly kleptoparasitises other seabirds. When displaying, males inflate their
bright red gular sac like a balloon and call to attract females, at the same time vibrating their
outstretched wings.

PLATE 8

1 Red-footed Booby   *Sula sula*

Piquero Patas Roj[

Length: 66–77 cm
Wingspan: 91–101 cm

Common but rather localised resident. Population estimated at *c.* 250,00[
pairs (including *c.* 140,000 pairs on Genovesa). Breeds throughout the yea[
nesting colonially in trees (the other boobies nest on the ground).

IDENTIFICATION: ADULT: Two colour forms: brown **1bf** an[
white **1wf** (*c.*95% of Galápagos population is of the brown form[
both forms have diagnostic combination of red feet and blue bi[
JUVENILE: Resembles brown form of adult but feet dark; distinguishe[
from other juvenile boobies by wholly brown underparts.

BEHAVIOUR: Courtship display, performed on branches near nes[
involves birds lifting their feet and waving them in the air. Partiall[
nocturnal, feeding well away from land. This perhaps explains wh[
it is less frequently encountered at sea than other boobies despite being the most numerou[
species in Galápagos.

2 Nazca Booby   *Sula granti*

Piquero Enmascarado; Piquero Blanc[

Length: 81–92 cm
Wingspan: *c.* 152 cm

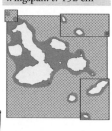

Common resident. Population *c.* 25,000–50,000 pairs. Breeds throughou[
the year, with colonies on different islands nesting at different times (e.[
eggs laid on Genovesa between August and November and on Español[
between November and February); nests on the ground. Formerly treated a[
a subspecies of Masked Booby *Sula dactylatra* but now afforded full specie[
status.

IDENTIFICATION: ADULT: Sexes alike, although females usuall[
slightly larger with duller bill; only black and white booby wit[
orange-yellow bill. JUVENILE: White underparts with distinct brow[
'bib', the white breast extending to form a narrow white collar.

BEHAVIOUR: Usually feeds well away from land.

e 3 Blue-footed Booby   *Sula nebouxii*

Piquero Patas Azule[

Length: 76–84 cm
Wingspan: *c.* 152 cm

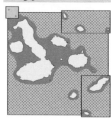

Common resident; endemic subspecies *excisa*. Population estimated a[
c. 10,000 pairs; breeds throughout the year, nesting on the ground.

IDENTIFICATION: ADULT: Sexes alike, although females have[
pigmented area around iris making pupil appear larger; blue fee[
diagnostic. JUVENILE: Resembles juvenile Nazca Booby but lacks well-
defined brown 'bib' of that species and shows white patch at base o[
hindneck and white rump.

VOICE: At breeding colony, males give a plaintive whistle wherea[
females and immatures give a hoarse "*quack*".

BEHAVIOUR: Courtship display, performed near nest, involve[
birds lifting their feet and waving them in the air. Usually feed[
close inshore.

4 Brown Booby   *Sula leucogaster*

Piquero Pardo

Length: 64–74 cm
Wingspan: 132–150 cm

Vagrant; one confirmed record of an adult. Occurs regularly in waters to the
north of Galápagos so possibly overlooked.

IDENTIFICATION: ADULT: Most likely to be confused with juvenile
Nazca Booby, from which distinguished in flight by yellow bill and lack of white collar.
JUVENILE: Resembles adult, but duller with underparts mottled brown.

PLATE 9

✓ ➊ Swallow-tailed Gull *Creagrus furcatus* Gaviota de Cola Bifurcad

E

Length: 51–58 cm
Wingspan: *c.* 130 cm

Common resident, especially in the easternmost islands. Populatio *c.* 10,000–15,000 pairs in over 50 colonies. Endemic, except for a sma colony on Malpelo Island off the west coast of Colombia. Breeds throughou the year, nesting in the shore zone.

IDENTIFICATION: Unmistakable; the only common whitish gu with a distinctive forked tail. ADULT: Upperparts and neck grey underparts white. In breeding plumage has dark grey head, large ey with red eye-ring, and black bill with pale base and tip. Non-breedin adults have white head with dark eye-patch. In flight shows distinctiv 'triangular' pattern of grey back and wing-coverts, white secondarie and black primaries. JUVENILE: Head and underparts white, with blac eye-patch and ear-spot; upperparts scaly brown and white.

VOICE: Gives a range of calls which may have a function in echolocation; most frequen alarm call is a rattle interspersed with a piercing "*pee*".

BEHAVIOUR: Feeds mostly nocturnally, usually several miles from land. Flight buoyan and tern-like.

✓ ➋ Lava Gull *Larus fuliginosus* Gavioto de Lava

E

Length: 51–55 cm
Wingspan: *c.* 130 cm

Widely distributed resident. The total world population occurs on Galápago and is fewer than 400 pairs. Conservation status: VULNERABLE. Breeds throughout the year, but mainly from May to October, nesting singly in the shore zone.

IDENTIFICATION: Unmistakable; the only all-dark gull, with heavy bill. ADULT: Blackish head with white eyelids, sooty-grey upperparts and breast, paler on belly. In flight, darker primaries contrast with rest of wing; rump and outer tail feathers whitish. JUVENILE: Dark chocolate-brown overall, except for pale rump. FIRST-WINTER/FIRST-SUMMER: Similar to juvenile, but has greyer tints to head and upperparts.

BEHAVIOUR: A tideline scavenger, rarely alighting on the sea.

➌ Pomarine Jaeger/Skua *Stercorarius pomarinus* Salteador

Length: 46–51 cm
Wingspan: 125–138 cm

Vagrant; few records, but probably overlooked as the waters off western South America are one of its main non-breeding ranges.

IDENTIFICATION: A smaller, slimmer version of South Polar Skua with distinctive broad tail-streamers, although these are lacking in juveniles and some adults. Wings show pale flashes at base of primaries. ADULT: Occurs in two colour phases: pale morph has whitish belly contrasting with otherwise dark brown plumage; dark morph is uniformly dark brown. Non-breeding adults show some barring on underparts in pale phase. JUVENILE: Underparts heavily barred brown and white. FIRST-WINTER: Like non-breeding adult, but has barring also on underwing in pale phase.

BEHAVIOUR: Similar to that of South Polar Skua.

➍ South Polar Skua *Catharacta maccormicki* Salteador Polar

Length: 50–55 cm
Wingspan: 130–140 cm

Vagrant, but probably under-recorded as it is a regular trans-equatorial migrant in the Pacific.

IDENTIFICATION: A large, heavy skua with conspicuous white wing-flashes at the base of the primaries. Occurs in two colour phases: pale morph shows obvious contrast between light grey head and underparts, and dark brown upperparts; dark morph is uniformly dark brown except for pale, brown-streaked collar.

BEHAVIOUR: Kleptoparasitises other seabirds, displaying surprising agility for its bulk. Normal flight is direct, low over the water and with heavy, deliberate wingbeats.

PLATE 1(

1 Laughing Gull   *Larus atricilla*

Gaviota Reidor

Length: 39–46 cm
Wingspan: 102–107 cm

Regular migrant, mostly from October to May. Breeds in North and Centra America and the Caribbean, some birds migrating to winter along the we coast of South America.

IDENTIFICATION: Similar to Franklin's Gull but slightly larger and generally darker o the upperparts with longer bill, which has a pronounced droop, flatter crown and longer leg ADULT BREEDING: Similar to Franklin's, but in flight rather more pointed wings show predominantly black wing-tips. ADULT NON-BREEDING: As adult breeding but pale-headed lacking the partial hood of Franklin's. FIRST-WINTER: As in adult non-breeding, hood les pronounced than in equivalent plumage of Franklin's, though displays indistinct black marking around ear-coverts. Underparts dusky compared with Franklin's, and in flight the black sub terminal tail-band is unbroken, extending to the edge of the tail.

2 Franklin's Gull   *Larus pipixcan*

Gaviota de Frankli

Length: 32–38 cm
Wingspan: 87–91 cm

Regular migrant, mostly from October to May. Breeds in North America spending the northern winter on the west coast of South America.

IDENTIFICATION: A smallish gull. ADULT BREEDING: White underparts and tail wit medium-grey mantle and upperwings. Black hood with bold white crescents above and below eye. Bill dark red. In flight shows equal amount of black and white at wing-tip. ADULT NON BREEDING: Differs from breeding adult in much reduced hood, with black only around and behind the eye. FIRST-WINTER: Saddle dark grey; rest of upperparts and upperwings grey brown; partial hood as in non-breeding adult. In flight, white tail has black subterminal band which does not extend to the edge of the tail as the outermost tail feathers are white.

3 Kelp Gull   *Larus dominicanus*

Gaviota Dominicana

Length: 54–65 cm
Wingspan: 128–142 cm

Vagrant: few records.

IDENTIFICATION: A large, rather heavily-built gull with a stout bill. ADULT: All-white apart from black mantle and upperwings. In flight shows white trailing-edge to wings and small white 'windows' at wing-tips. Bill yellow with red spot; legs greenish-yellow. JUVENILE: Dark brown with buff edges to flight and tail feathers; bill dark. FIRST-WINTER/FIRST-SUMMER: Similar to juvenile but with paler head, greyer upperparts and whiter underparts. SECOND-WINTER/SECOND-SUMMER: Head and underparts white, faintly streaked with brown; wings and mantle black with grey-brown coverts; bill dull yellow with indistinct orange spot; legs dull yellow.

VOICE: A harsh, repeated "*kree–ok*".

① Red-billed Tropicbird *Phaethon aethereus*

Ave Tropica

Length: 90–105 cm
Wingspan: 99–106 cm

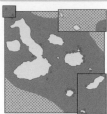

Common resident; population estimated to be a few thousand pairs in abou 30 colonies. Nests colonially in rocky crevices or on bare ground, breedin; throughout the year.

IDENTIFICATION: ADULT: Large, mainly white seabird with bright red bill, yellow legs and very long, white central tail-streamer (46–56 cm). Broad black eye-stripe extends back towards the nape Upperparts have fine grey barring. In flight shows black primaries JUVENILE: Similar to adult, but lacks the tail-streamers, has a yellow bill, and the black eye-stripes meet at the nape.

VOICE: A loud, shrill rattle, "*kree-kree-kree*".

BEHAVIOUR: Feeds by hovering and plunge-diving, usually fa from land. Flight is graceful, alternating fluttering wingbeats with gliding. Often seen resting on the sea with the tail raised.

② Common Noddy *Anous stolidus*

Gaviotín Cabeza Blanca

Length: 38–45 cm
Wingspan: 75–86 cm

Common resident; endemic subspecies *galapagensis*. Population estimated to be a few thousand pairs. Breeds throughout the archipelago in small colonies on sea cliffs just above the tideline.

IDENTIFICATION: The only entirely dark tern, with typical pointed wings, and wedge-shaped tail. Smaller and more elegant than Lava Gull. ADULT: Dark brown overall, except for whitish-grey forehead and crown, and white eyelids. In flight shows pale wing-bar on upperwing and dusky-grey centre to underwing. JUVENILE: Similar to adult, but lacks contrasting cap.

VOICE: Usually silent, but occasionally emits a low guttural growl.

BEHAVIOUR: Hovers above sea, often in large groups, sometimes accompanying pelicans, which may be used as a perch! Generally flies low over the sea, picking food from the surface; does not plunge-dive.

③ Sooty Tern *Sterna fuscata*

Gaviotín Negro

Length: 43–45 cm
Wingspan: 86–94 cm

Resident; breeding only on Darwin, where it is present in very large numbers. Seen only infrequently elsewhere in Galápagos.

IDENTIFICATION: Unmistakable; the only tern with contrasting black and white plumage. ADULT: Upperparts entirely black apart from white triangle on forehead; underparts white. In flight shows white outer tail-streamers. JUVENILE: Mostly sooty-brown, finely spotted white above, dusky below with paler belly.

VOICE: Noisy, with loud "*ker–wacky–wack*" call at breeding colony.

BEHAVIOUR: Very buoyant flight. Often seen feeding in large flocks, snatching food from the surface of the sea; does not plunge-dive.

① Black Tern   *Chlidonias niger*   Gaviota Negr

Length: 23–28 cm
Wingspan: 58–66 cm

Vagrant; one record of a dead immature bird.

IDENTIFICATION: A small, dark tern with a thin, dark red bi
ADULT BREEDING: Black head, neck and underparts, except for white vent and undertai
Upperwing dark grey; tail and underwing pale grey. ADULT NON-BREEDING: Pale grey abov
white below, with dark shoulder-patch. Head white with black on crown extending to nap
ear-coverts and through eye. FIRST-WINTER: Similar to non-breeding adult, but upperpar
show brownish tinges, especially on upperwing in flight.

BEHAVIOUR: Migrants usually feed in coastal waters, dipping down to take food from th
surface, frequently hovering. Flight graceful and buoyant.

② Common Tern   *Sterna hirundo*   Gaviotín Comú

Length: 32–39 cm
Wingspan: 72–83 cm

Regular migrant in small numbers. Breeds in the northern hemisphere, th
North American population wintering off Central and South America. Mos
Galápagos records have been during the northern winter.

IDENTIFICATION: A medium-sized tern, pale grey above and white below with a blac
cap and short red legs. ADULT BREEDING: Black cap extends from bill to nape. Bill red with
black tip. Long, forked tail reaches wing-tip when perched. In flight shows dark outer web
to primaries and dark trailing-edge to primaries on underwing. ADULT NON-BREEDING: A
adult breeding but forehead white and bill all black. FIRST-WINTER: Like non-breeding adult
but has prominent dark carpal bar and dusky trailing-edge to secondaries.

BEHAVIOUR: A typical tern with light and buoyant flight. Feeds by plunge-diving from th
air and surface-dipping.

③ Royal Tern   *Sterna maxima*   Gaviotín Rea

Length: 45–51 cm
Wingspan: 100–135 cm

Regular migrant in small numbers, most records having been from the south
of Isabela and Santa Cruz from January to March. Breeds in North America
the Caribbean and northern South America, some birds migrating to winter
along the north-west coast of South America.

IDENTIFICATION: A large, stocky tern with a heavy, orange bill and black legs. ADULT
BREEDING: Upperparts pale grey; underparts white. Black crown, extending from bill to nape.
In flight shows white forked tail and dark trailing-edge to primaries on underwing. ADULT
NON-BREEDING: Similar to adult breeding but lores, forehead and forecrown white, merging
with variable black spotting on crown, to shaggy black nape. FIRST-WINTER: Resembles non-
breeding adult, but has dark brown markings on wings.

VOICE: A loud, harsh "*kee–reer*"

BEHAVIOUR: Could be mistaken for a gull due to its large size and powerful flight. When
feeding, flies relatively high over the water and plunge-dives or surface-dips.

④ White Tern   *Gygis alba*   Gaviotín Blanco

Length: 25–30 cm
Wingspan: 76–80 cm

Vagrant; very few records. Breeds on tropical islands throughout the world.

IDENTIFICATION: Unmistakable; a small, very dainty, white tern
with slightly upturned black bill and blue-grey legs. Graceful in flight, with almost translucent
wings and a shallow-forked tail. ADULT: Plumage entirely white except for black eye-ring.
JUVENILE: Similar to adult but back brownish-grey and shows dark smudges behind the eye
and on the nape.

BEHAVIOUR: A tame and inquisitive species. Feeds by diving onto the surface of the sea.

❶ Blue-winged Teal   *Anas discors*   Cerceta Aliaz

Length: 39 cm

Vagrant; few records, mainly from coastal lagoons but also from ponds in highlands, mostly during December to April.

IDENTIFICATION: A small, mainly brown dabbling duck with pale yellowish legs and fe distinctive blue forewing and pale underwing in flight. MALE: Unmistakable in breeding pluma with slate-grey head and large white facial crescent between eye and bill, and black undertail-cove with white 'spot' on rear flanks. FEMALE/IMMATURE/ECLIPSE MALE: Brown with spotted underpa and faintly streaked face and neck with small, whitish loral spot; bill dark grey.

❷ Masked Duck   *Oxyura dominica*   Malvasía Enmascara

Length: 30–36 cm

Vagrant; one record of a pair with a duckling on El Junco Lake, San Cristóbal December 1994.

IDENTIFICATION: A small, compact diving duck, readily told in flight by conspicuous whi patch across centre of wing, and on water by long, pointed tail which is often raised at about 45 MALE: Unmistakable in breeding plumage, with rufous body mottled with dark brown and blac black face; and blue bill tipped black. FEMALE/IMMATURE/ECLIPSE MALE: Brown with mottle underparts and buff head with black stripes through and below eye and on crown; bill bluish.

❸ Pied-billed Grebe   *Podilymbus podiceps*   Somormuj

Length: 34 cm

Vagrant; recorded regularly, mainly on brackish lagoons, particularly on S Cristóbal, and may have bred.

IDENTIFICATION: The only grebe recorded. Told from other waterbirds by aquatic habit frequently diving; small size; short-necked and large-headed appearance with very short tail; unifor grey-brown plumage with white undertail-coverts; and stout, pale bill which, during the breedin season, has a vertical black band.

❹ American Coot   *Fulica americana*   Focha American

Length: 34–43 cm

Vagrant; first recorded in 1999. Breeds mainly in North and Central America an the West Indies, migrating to winter usually only as far south as Panamá.

IDENTIFICATION: A largish, plump rail, distinguished from similar Common Gallinule (se Plate 14) by lack of white flank-stripe and colour of frontal shield. ADULT: Plumage entirely slate-gre apart from white outermost undertail-coverts. White bill and frontal shield diagnostic, the bill havin a broken reddish subterminal band.

❺ White-cheeked (or Galápagos) Pintail   *Anas bahamensis*   Patill

Length: 46 cm

Fairly common resident; endemic subspecies *galapagensis*. Breeds opportunisticall throughout the year when conditions are favourable, nesting near the ground i waterside vegetation.

IDENTIFICATION: Unmistakable: a medium-sized, slender dabblin duck with white cheeks and throat, and black bill with red sides to base In flight appears slender with long, pointed tail and conspicuous whit cheeks and throat.

VOICE: Usually silent, although when calling, males make a low whistl and females utter a weak, descending series of quacks.

BEHAVIOUR: Usually found singly or in pairs or small groups near open wetlands, both near the coast and in the highlands. Usually feeds b dabbling and up-ending, although regularly dives for food in deep water.

❻ Black-bellied Whistling Duck   *Dendrocygna autumnalis*   Guirir

Length: 48–58 cm

Vagrant; one record from Isabela.

IDENTIFICATION: ADULT: Unmistakable: a medium-sized, mainly rufous-brown duck with red bill and legs, black underparts, grey head and upper-neck, and prominent white wing-flash. IMMATURE: Plumage similar to adult but bill and legs blue-grey.

PLATE 1

❶ **Paint-billed Crake** *Neocrex erythrops* Gallare

Length: 18–20 cm

Scarce resident; found principally in agricultural areas, favouring dense, wetlar vegetation although also occurs in damp woodlands and in dry pastures and thicke Breeds from November to February.

IDENTIFICATION: A small, rather dark rail. ADULT: Upperparts oliv brown; underparts slate-grey with narrow white vertical bars on flank bill pale green with bright red base and black tip; legs and feet red. JUVENIL Similar to adult but plumage generally duller and paler with dark b lacking the red base.

VOICE: Song is a long, descending and gradually accelerating series staccato notes followed by a few short churrs. Call is a frog-like "*qurrk* either given singly or in a series.

BEHAVIOUR: Mainly crepuscular and very secretive. Rarely flies unless flushed.

Ⓔ ❷ **Galápagos Rail** *Laterallus spilonotus* Pacha

Length: 15–16 cm

Uncommon resident; breeds from September to April. Population declining an range becoming increasingly restricted; now virtually confined to the highlanc where it inhabits dense grassy vegetation, thickets and forests. Conservation Statu NEAR-THREATENED.

IDENTIFICATION: A tiny, dark, short-winged rail. ADULT: Upperpart dark chocolate-brown, finely spotted with white; underparts dark slate grey to greyish-brown; eyes scarlet; legs and feet brown. JUVENILE: Simila to adult but plumage generally duller and slightly paler and lacks th white spots on the upperparts; eye dark.

VOICE: Gives a range of calls but typically a rapid "*chi–chi–chi–chirroo*"

BEHAVIOUR: Rather furtive, although can be very tame and sometime inquisitive. Flight very weak and runs rather than flies when disturbed.

❸ **Sora (Rail)** *Porzana carolina* Polluela Norteña

Length: 19–25 cm

Vagrant; three records, all of birds found dead. Normally favours freshwater, brackish and coastal marshes with emergent vegetation.

IDENTIFICATION: A small, plump rail with 'stubby' yellow or greenish-yellow bill. ADULT: Shows obvious black face 'mask'. JUVENILE/IMMATURE: Similar to adult but duller.

✓ ❹ **Common Gallinule (or Moorhen)** *Gallinula chloropus* Gallinula

Length: 30–38 cm

Resident, mainly inhabiting brackish lagoons; breeds mostly from May to October.

IDENTIFICATION: A medium-sized, dark gallinule. ADULT: Generally dark blackish-grey, with distinctive broken white line along flanks, and white undertail-coverts; frontal shield and base of bill bright red and tip of bill yellow; legs and feet yellow, with upper part of tibia orange. JUVENILE: Much duller than adult, with pale brown underparts and greenish-brown bill, lacking the frontal shield; legs and feet olive-grey.

VOICE: Typically an explosive "*krrruk*", but also gives a range of sharp "*kik*" and "*kark*" calls.

BEHAVIOUR: Distinctive, jerky head movement when swimming.

❺ **American Purple Gallinule** *Porphyrio martinica* Gallito Azul

Length: 27–36 cm

Vagrant; few records.

IDENTIFICATION: A medium-sized, rather slender gallinule. ADULT: Head, neck and underparts iridescent purplish-blue, with white undertail-coverts; upperparts olive-green; frontal shield pale blue; bill bright red at base with yellow tip; legs and feet yellow. JUVENILE: Much duller than adult with pale brown underparts; yellow-green bill with pale tip, lacking the frontal shield; and yellowish or brownish legs and feet.

PLATE 1

1 Greater Flamingo *Phoenicopterus ruber*
Flamenc

Length: *c.* 120 cm
Wingspan: *c.* 140 cm

Resident; sometimes treated as an endemic subspecies *glyphorhynchu*
Population estimated at between 400–500 birds. Breeds in small coloni
from July to March, building mud nests in saltwater lagoons.

IDENTIFICATION: Unmistakable; a very large, long-necked, pin
bird with distinctive 'kinked' bill. Neck extended in flight. ADUL
Wholly pink with conspicuous black flight feathers in fligh
IMMATURE: As adult but plumage whitish.

VOICE: A series of rather goose-like "*ah–ah–ah* ..." calls.

BEHAVIOUR: Feeds in small groups in saltwater lagoon
Occasionally seen at sea flying from one island to another.

e 2 Great Blue Heron *Ardea herodias*
Garza Moren

Length: *c.* 95 cm
Wingspan: *c.* 175 cm

Resident, endemic subspecies *cognata*. Found in small numbers in coast
areas, particularly near lagoons. Breeds throughout the year, nesting singl
close to the shore, usually in mangroves.

IDENTIFICATION: Unmistakable: a very large, long-necked, gre
heron. ADULT: Upperparts pale grey; sides of neck violaceous-grey
head white with broad black line along side of crown extending t
long crest; underparts mostly white with black spots on breast; leg
and bill yellow, becoming pinkish during the breeding season
IMMATURE: Grey-brown with dark bill.

VOICE: Gives a harsh "*kraak*" in flight and a soft "*frawnk*" i
disturbed near the nest.

BEHAVIOUR: Usually found singly. Sometimes flies with neck outstretched.

3 Tricolored Heron *Egretta tricolor*
Garza Tricolo

Length: *c.* 60 cm
Wingspan: *c.* 100 cm

Vagrant; recorded once.

IDENTIFICATION: A medium-sized heron with long, slende
neck and relatively short legs. ADULT: Blue-grey upperparts, head and neck with white lin
running down front of neck; underparts and, in flight, underwing-coverts white; bill grey
with dark tip; lores yellow. JUVENILE: Similar to adult but sides of head and neck rufous and
mantle and wing-coverts tinged rufous.

4 Little Blue Heron *Egretta caerulea*
Garza Azul

Length: *c.* 65 cm
Wingspan: *c.* 100 cm

Vagrant; recorded once, in 1998.

IDENTIFICATION: A medium-sized, long-necked heron. ADULT:
Uniform slate-blue plumage with purple or cinnamon wash to head and neck; bill blue-grey
with dark tip; legs and feet black. JUVENILE/IMMATURE: White, blotched with variable amounts
of blue-grey (see Plate 16).

PLATE 1(

① Cattle Egret   *Bulbulcus ibis*   Garza (del Ganado) Bueyer:

Length: *c.* 50 cm
Wingspan: *c.* 90 cm

Resident. First recorded in 1964 but breeding not proved until 1986; now well established, particularly in agricultural areas on the inhabited islands Nests colonially in bushes or trees.

IDENTIFICATION: A medium-sized, short-necked and rathe compact white heron with yellow bill and relatively short, dark green legs and feet which can appear black at a distance. In breeding plumage, develops rufous-buff plumes on the crown, nape, mantle and upper breast, and the legs become yellow and, for a short while bright red.

VOICE: At breeding colony typically gives a harsh "*rick-rack*" call although also has a range of other harsh calls including "*raa*", "*kraa*" "*thonk*" and "*kok*".

BEHAVIOUR: Usually found in loose flocks in agricultural areas feeding on invertebrate: disturbed by livestock or tortoises, although sometimes seen near the coast, particularly around lagoons.

② Little Blue Heron   *Egretta caerulea*   Garza Azul

Length: *c.* 60 cm
Wingspan: *c.* 100 cm

Vagrant; recorded once, in 1998.

IDENTIFICATION: A medium-sized, long-necked heron. ADULT: Slate-blue (see Plate 15). JUVENILE: White with yellowish-green legs and feet and grey or pinkish bill with dark tip; lores usually grey. Dark tips to primaries diagnostic. IMMATURE: White plumage blotched with blue-grey; white feathers gradually replaced with blue-grey as bird matures.

③ Snowy Egret   *Egretta thula*   Garcita Blanca

Length: *c.* 60 cm
Wingspan: *c.* 100 cm

Vagrant; recorded most years in small numbers, particularly in coastal areas.

IDENTIFICATION: A medium-sized, long-necked, white heron readily identified by combination of black bill with yellow lores and black legs with yellow feet.

④ Great Egret   *Ardea alba*   Garza Blanca

Length: *c.* 80 cm
Wingspan: *c.* 150 cm

Resident. Found in small numbers in coastal areas, particularly near lagoons, and occasionally in open areas in the highlands. Nests colonially, usually in mangroves.

IDENTIFICATION: A large, long-necked, white heron with yellow bill and black legs and feet, the upper part of the leg sometimes becoming yellowish during the breeding season. Gape extending behind eye is diagnostic.

VOICE: At breeding colony gives a soft "*frawnk*" or a deep "*rhaa*" or "*rhoo*"; in flight sometimes gives a low-pitched "*krraak*".

BEHAVIOUR: Usually found singly away from the breeding colony.

PLATE 17

1 Yellow-crowned Night-heron *Nycticorax violaceus*   Guaque; Garza Nocturna

Length: *c.* 60 cm
Wingspan: *c.*100 cm

Fairly common resident; endemic subspecies *pauper*. Mainly found in coastal areas, breeding throughout the year and nesting close to the ground amongst mangrove roots or rocks and sometimes in caves.

IDENTIFICATION: A medium-sized, squat heron, with legs protruding beyond the tip of the tail in flight. ADULT: Unmistakable: generally dark grey with pale fringes to wing and back feathers; black head with white cheeks and yellow crown; legs yellow and bill black. IMMATURE: Similar to adult but plumage brown rather than grey and head pattern indistinct. JUVENILE: Dark brown with white or buff spots and streaks on upperparts, and creamy-white and brown streaks on underparts; lacks the head markings of adult. Underwings greyish.

VOICE: A range of calls but typically an often-repeated, rather high-pitched "*kwok*"; in flight gives a "*scalp*".

BEHAVIOUR: Mainly nocturnal, often seen flying from roost sites at dusk to feed; usually a solitary feeder but flocks can sometimes be found around coastal lagoons.

2 Black-crowned Night-heron *Nycticorax nycticorax*   Garza Nocturna

Length: *c.* 60 cm
Wingspan: *c.* 100 cm

Vagrant; one record of an immature bird on Santa Cruz in 1971.

IDENTIFICATION: Similar in size and shape to Yellow-crowned Night-heron but posture typically rather more hunched with neck tucked into shoulders; in flight only tips of toes project beyond the tip of the tail. ADULT: Unmistakable: underparts, neck and face white; wings uniform pale grey; back and crown black; legs yellow and bill black. JUVENILE: Very similar to juvenile Yellow-crowned, but paler with more numerous and larger, buffy-white spots and streaks on upperparts, and darker streaking on underparts. Underwings whitish.

VOICE: Flight call is a hoarse, low-pitched "*kwark*".

3 Striated Heron *Butorides striatus*   Garza Verde

Length: 35 cm
Wingspan: 63 cm

Uncommon resident, probably breeding throughout the year when conditions are suitable; nests usually solitary.

IDENTIFICATION: A very small, compact heron. ADULT: Back and wings greenish-grey, with wing coverts fringed brown; cheeks and sides of neck pale grey, contrasting with black crown; underparts pale grey with white line down front of neck and often with brownish wash to sides of lower neck and upper breast. Legs dull greyish-yellow, becoming bright yellow when breeding. JUVENILE: Indistinguishable from juvenile Lava Heron until it begins to attain adult plumage.

VOICE: Alarm call is a sharp "*keyow*", indistinguishable from Lava Heron.

BEHAVIOUR: Found in similar habitats to Lava Heron, but more often in freshwater habitats.

E 4 Lava (or Galápagos) Heron *Butorides sundevalli*   Garza de Lava

Length: 35 cm
Wingspan: 63 cm

Common resident; considered by some authorities to be a subspecies of Striated Heron. Mainly found in coastal areas, breeding throughout the year when conditions are suitable; nests usually solitary, built low down in mangroves or under rocks.

IDENTIFICATION: A very small, compact heron. ADULT: Unmistakable: wholly blackish-grey with dull greyish-orange legs which become bright yellow or orange when breeding. JUVENILE: Predominantly brown, with whitish tips to wing-coverts and pale fringes to wing feathers; breast pale with brown streaking; crown dark; back uniform grey; cheeks and sides of neck washed brown. Indistinguishable from juvenile Striated Heron until it begins to attain adult plumage.

VOICE: Alarm call is a sharp "*keyow*".

BEHAVIOUR: Feeds in the shore zone.

PLATE 18

1 Black-necked Stilt *Himantopus mexicanus*

Tero Rea[

Length: 35–40 cm

Uncommon resident. Found in saline and freshwater habitats throughou[the islands, breeding from December to June. This New World form [sometimes considered to be the same species as the Black-winged Sti[*H. himantopus* of the Old World.

IDENTIFICATION: A medium-sized, unmistakable wader wit[black wings, mantle, hindneck and crown; white underparts, rum[and tail; and extremely long, pink legs which project well beyon[the tip of the tail in flight. Sexes alike although females have brow[cast to mantle.

VOICE: Gives a range of high-pitched piping calls, including "*kik kik-kik...*", "*kek*" and "*kee-ack*".

2 Black Turnstone *Arenaria melanocephala*

Vuelvepiedras Negr[

Length: 22–25 cm

Vagrant; one record from the highlands of San Cristóbal.

IDENTIFICATION: Size and shape as Ruddy Turnstone with similar wing-pattern in flight. Differs from Ruddy in having black head, throat and breast (although breeding adult has white spot at base of bill), and darker, greyish legs.

3 Surfbird *Aphriza virgata*

Chorlito de Rompientes

Length: 23–26 cm

Scarce migrant, mainly during the northern winter; confined to rocky coasts.

IDENTIFICATION: A smallish, rather stocky wader, similar in size and shape to the turnstones although bill blunt-tipped and not upturned; in flight lacks the turnstones' white rump. ADULT BREEDING: Head, neck and breast heavily streaked black extending to black chevrons on flanks; golden-buff markings on scapulars. ADULT NON-BREEDING AND JUVENILE: Similar to Black Turnstone but throat white and legs yellow; shows some black markings on flanks.

4 Ruddy Turnstone *Arenaria interpres*

Vuelve Piedras

Length: 21–26 cm

A common migrant, present throughout the year although most numerous from September to March; confined to the coast.

IDENTIFICATION: A smallish, rather stocky wader with relatively short, orange legs and short, slightly upturned bill. In flight shows distinctive variegated wing-pattern. ADULT BREEDING: Back and wings rufous; distinctive black and white facial pattern and black breast-band contrasting with unmarked white underparts; throat white. ADULT NON-BREEDING AND JUVENILE: Similar to, but duller than, breeding adult, with less rufous on upperparts and less distinct head-pattern.

VOICE: A staccato rattle "*tuk, tuk–e–tuk–tuk*".

BEHAVIOUR: Usually occurs in small flocks.

5 American Oystercatcher *Haematopus palliatus*

Ostrero

Length: 40–44 cm

Uncommon resident; endemic subspecies *galapagensis*. Population probably numbers around 200 pairs. Confined to rocky shores, sandy beaches and coastal lagoons; breeds mainly from October and March.

IDENTIFICATION: A large, unmistakable wader with black head and neck, dark brown upperparts, white underparts, long orange bill and rather short, pink legs. In flight shows prominent white wing-bar and white rump contrasting with black tail.

VOICE: A distinctive, shrill piping "*kleet*".

BEHAVIOUR: Usually found in pairs.

PLATE 1

1 Wilson's Plover   *Charadrius wilsonia*   Chorlo de Wilso

Length: 20 cm   Vagrant; one record of three birds on Floreana in May 1969.

IDENTIFICATION: Slightly larger than Semipalmated Plover, with
longer, heavier, black bill. ADULT BREEDING: Upperparts dark grey-brown, underparts white wi
broad black breast-band (brown in female). Brown and white head-pattern includes (male onl
black forecrown. ADULT NON-BREEDING/JUVENILE/FIRST-WINTER: Resembles breeding female, but lac
any rufous tinges; upperparts more scaly in juvenile.
VOICE: A high-pitched, short whistle: "*Wheet*".

2 Killdeer   *Charadrius vociferus*   Playero Grito

Length: 27 cm   Vagrant; one record from Isabela in February 1971.

IDENTIFICATION: A distinctive, medium-sized plover with blac
double breast-band and long tail. In flight shows bright chestnut rump and broad white wing-ba
ADULT: Upperparts dark brown, with rufous fringes in non-breeding plumage. JUVENILE: Similar
non-breeding adult, but has paler fringes.
VOICE: A very loud, penetrating "*Kill–dee*", uttered repeatedly.

3 Semipalmated Plover   *Charadrius semipalmatus*   Chorlitej

Length: 18 cm   Fairly common migrant, most numerous from August to May, although prese
                     throughout the year, occurring in the shore zone.

IDENTIFICATION: A small plover, brown above with white collar, and white below with da
breast-band. Conspicuous white wing-bar in flight. ADULT BREEDING: Contrasting black and whit
head-pattern, black breast-band. Bill orange with black tip. ADULT NON-BREEDING: Black parts c
breeding plumage replaced by dark brown and bill darker. JUVENILE: Similar to non-breeding adul
but with buff fringes to upperparts.
VOICE: A plaintive, rising "*Chew–ee*", accented on the second syllable.
BEHAVIOUR: Usually found on sandy shorelines, but occasionally occurs in the highlands besid
freshwater pools.

4 Pacific Golden Plover   *Pluvialis fulva*   Playero Dorado Siberianc

Length: 27 cm   Vagrant; one record.

IDENTIFICATION: Extremely similar to the American Golden Plove
in all plumages. Pacific has slightly longer legs, shorter wings and heavier bill. ADULT BREEDING
White neck-stripe extends to side of belly, with smudgy-grey coloration. ADULT NON-BREEDING
JUVENILE/FIRST-WINTER: More golden, with less grey tones to upperparts and breast than Americar
Golden Plover; supercilium buff. **VOICE:** A sharp "*Chu–it*".

5 American Golden Plover   *Pluvialis dominica*   Playero Dorado Americana

Length: 27 cm   Vagrant; one record from Santa Cruz.

IDENTIFICATION: Both this and the very similar Pacific Golden Plover
are daintier versions of the Black-bellied Plover, with longer legs and slimmer bill. ADULT BREEDING
Underparts jet-black. Upperparts spangled gold and black, with white neck-stripe. ADULT NON-
BREEDING/JUVENILE/FIRST-WINTER: Rather plain with dark-grey upperparts flecked with gold; prominent
white supercilium.
VOICE: A clear, melancholy whistle, either a single or a double note.

6 Black-bellied (or Grey) Plover   *Pluvialis squatarola*   Playero Cabezón

Length: 29 cm   Regular migrant, recorded throughout the year. Occurs in small numbers, mostly
                     on sandy beaches.

IDENTIFICATION: A plump, medium-sized wader with short, stout bill. ADULT BREEDING:
Underparts jet-black from face to belly, separated from spangled grey and silver upperparts by broad
white stripe on side of neck. ADULT NON-BREEDING/JUVENILE/FIRST-WINTER: Grey-brown upperparts,
spotted white (buff in juvenile); pale-brown underparts. Distinctive black 'armpits' in flight.
VOICE: Flight call is a loud, three-note whistle: "*Pee–oo–wee*".

PLATE 2(

1 **Willet** *Catoptrophorus semipalmatus*

Chorlitej

Length: 33–41 cm

Regular migrant, occurring in the shore zone.

IDENTIFICATION: A large, plump wader with a heavy straigh bill and longish blue-grey legs. Greyish overall, but has a striking black and white pattern o the open wing. ADULT BREEDING: Upperparts greyish-brown, lightly speckled black. Underpar white, with faint barring on breast and flanks. ADULT NON-BREEDING: Uniformly pale gre above and white below, with grey wash to breast. JUVENILE: Similar to non-breeding adult bu upperparts slightly browner with buff fringes.

VOICE: Gives repeated shrill alarm call, "*weet, weet*" and loud triple flight call, "*will–will–willet*'

2 **Short-billed Dowitcher** *Limnodromus griseus*

Agujeta Piquicort

Length: 25–29 cm

Uncommon migrant, occurring in the shore zone.

IDENTIFICATION: A medium-sized wader with short, greenis legs and a long, snipe-like bill. In flight shows a whitish back. ADULT BREEDING: Upperpart dark brown with narrow rufous fringes. Underparts rufous with whitish belly, barred an spotted on flanks. Tail finely barred. ADULT NON-BREEDING: Grey above and on breast; pal underparts. JUVENILE: Brighter than adult, lacking barring on underparts. Crown an upperparts broadly edged reddish-buff.

VOICE: A soft and rapid "*tu–tu–tu*".

3 **Hudsonian Godwit** *Limosa haemastica*

Aguja Caf

Length: 36–42 cm

Vagrant; one record.

IDENTIFICATION: Smaller and shorter-legged than the Marble Godwit, with a proportionately shorter, thinner bill, also slightly upturned. In flight, show. dark wing-linings, narrow white wing-bar, and black and white tail. ADULT BREEDING: Femal is paler than male, with white blotching on underparts. ADULT NON-BREEDING: Dark grey upperparts and breast, pale belly and white supercilium. JUVENILE: Brownish-buff overall with pale fringes on upperparts.

VOICE: A shrill, high-pitched "*whit*", often repeated.

4 **Whimbrel** *Numenius phaeopus*

Zarapito

Length: 40–46 cm

Regular migrant, found on rocky shores and lagoons; also around pools and in grassy areas in the highlands..

IDENTIFICATION: Unmistakable: the only large wader with a long, downcurved bill. Long-legged, with conspicuously striped crown. ADULT: Upperparts grey-brown; underparts paler, with fine streaking on breast. JUVENILE: Upperparts slightly darker than in adult, with buff spots and fringes on flight feathers.

VOICE: A very characteristic and far-carrying six or seven note trill: "*tu–tu–tu–tu–tu–tu*".

BEHAVIOUR: Found on rocky shores and lagoons; also around pools and in grassy areas in the highlands.

5 **Marbled Godwit** *Limosa fedoa*

Aguja Canela

Length: 42–48 cm

Scarce migrant, mainly between October and March, and found in the shore zone.

IDENTIFICATION: A large, richly mottled, brown wader with long legs and long, slightly uptilted bill. In flight shows cinnamon wing-bar. ADULT BREEDING: Upperparts speckled black and cinnamon; underparts pale chestnut, lightly streaked and barred dark brown. Face whitish, with dark crown and lores. ADULT NON-BREEDING/JUVENILE/FIRST-WINTER: Paler overall than adult breeding, lacking streaks and bars on underparts; juvenile has buff fringes to mantle feathers.

VOICE: A loud, harsh "*ker–whit*".

PLATE 2

1 **Lesser Yellowlegs** *Tringa flavipes* Chorlo Chic

Length: 23–25 cm | Regular migrant, mostly between October and March, occurring in the sho zone.

IDENTIFICATION: Similar in structure to Greater Yellowlegs, but smaller and daintie with thinner, proportionately shorter bill (shorter than length of tarsus). ADULT BREEDIN Differs from Greater in having largely unmarked flanks. ADULT NON-BREEDING/JUVENIL Similar plumage to Greater, but less spotted on mantle and with finer breast streaking.

VOICE: A high-pitched "*tew–tew*", shorter and quieter than the call of Greater Yellowlegs

2 **Greater Yellowlegs** *Tringa melanoleuca* Chorlo Re

Length: 29–33 cm | Scarce migrant, occurring in the shore zone.

IDENTIFICATION: A fairly large, stocky wader with long, stou slightly upturned bill (equal to or longer than length of tarsus) and long, bright orang yellow legs. In flight shows white rump but no wing-bars. ADULT BREEDING: Head, throat an breast heavily streaked and flanks barred or with chevrons. Upperparts dark brown wit white spots. ADULT NON-BREEDING: Similar to adult breeding but upperparts grey with fine white spotting; underparts white, with grey wash to breast and flanks. JUVENILE: Resemble non-breeding adult, but darker on mantle, and shows clear contrast between streaky breas and white belly.

VOICE: A loud, descending series of three or more notes: "*tew–tew–tew*".

3 **Pectoral Sandpiper** *Calidris melanotos* Tin-Güír

Length: 20–23 cm | Vagrant; three records, two from the highlands of Santa Cruz and one from a coastal lagoon on Isabela during the northern winter.

IDENTIFICATION: A medium-sized wader with slightly decurved bill, strongly streake breast sharply contrasting with white belly and, typically, upright stance. ADULT MAL BREEDING: Breast blackish with white mottling. ADULT FEMALE BREEDING: Smaller than male with brown breast streaking. ADULT NON-BREEDING: Duller and greyer overall than in breeding plumage. JUVENILE: Upperparts brighter and more contrasting than adult, with buff an chestnut fringes, and more pronounced white supercilium.

4 **Buff-breasted Sandpiper** *Tryngites subruficollis* Correlimos Canelo

Length: 18–20 cm | Vagrant; one record.

IDENTIFICATION: A medium-sized wader, with short, dark bill and shortish orange-yellow legs. The strong buff tones of face and underparts, together with the black 'button' eye in the plain face, are distinctive at all ages. JUVENILE: distinguished from adult by more scaly appearance to the dark brown upperparts.

BEHAVIOUR: Favours areas of short grass.

5 **Stilt Sandpiper** *Micropalama himantopus* Correlinos Tarsilargo

Length: 18–23 cm | Vagrant; few records during the northern winter.

IDENTIFICATION: Resembles a long-legged *Calidris* sandpiper (see Plates 22 and 23), with a long, slightly decurved bill. In flight shows white rump but no wing-bars. ADULT BREEDING: Very distinctive, with rufous ear-coverts and crown-stripes, white supercilium and strongly barred underparts. ADULT NON-BREEDING: Greyer above and paler below than in adult breeding, lacking any strong markings. JUVENILE: Upperparts dark brown with rufous fringes on mantle and pale edges to wing-coverts and tertials.

PLATE 22

1 Red Knot *Calidris canutus*

Playero Pecho Rufo

Length: 23–25 cm

Vagrant; one record from Floreana in 1969.

IDENTIFICATION: A dumpy, medium-sized wader, with a fairly short, straight bill and short, greenish legs. In flight shows pale grey rump. ADULT BREEDING: Brick-red underparts and a mixture of black and chestnut on upperparts. ADULT NON-BREEDING: Underparts white with grey wash on breast; upperparts grey with narrow white fringes to feathers. JUVENILE: Like non-breeding adult, but has a buffish tinge to the underparts and broader pale fringes to feathers on upperparts.

2 Sanderling *Calidris alba*

Playero Común

Length: 20–21 cm

Common migrant, particularly during the northern winter, confined to the shore zone.

IDENTIFICATION: A very active small wader, distinctively pale in non-breeding plumage, with short black bill and legs. Runs energetically along the tideline. ADULT BREEDING: Head, mantle and breast rusty-red; underparts white. ADULT NON-BREEDING/FIRST-WINTER: Very pale grey above, with black shoulder-patch, and pure white below. JUVENILE: Mantle chequered black and white; creamy wash on breast and flanks.

VOICE: A sharp "*krit*", often repeated.

3 Spotted Sandpiper *Actitis macularia*

Correlino

Length: 18–20 cm

Regular migrant; most records during the northern winter from the shore zone.

IDENTIFICATION: A small, active wader with constantly bobbing tail, shortish olive-yellow legs and medium-length bill; flies on stiff, bowed wings with a curious stuttering action. ADULT BREEDING: Strikingly plumaged, with white underparts liberally sprinkled with black spots, and finely barred, olive-brown upperparts. Bill orange with black tip. ADULT NON-BREEDING/JUVENILE: Plain upperparts and unspotted white underparts, extending up side of breast to form a white shoulder patch.

VOICE: High, piping "*peet–weet–weet*" notes, usually given in flight.

4 Wandering Tattler *Heteroscelus incanus*

Errante

Length: 26–29 cm

Common migrant; found throughout the year on rocky coasts.

IDENTIFICATION: A rather plain, medium-sized wader, with a stout, straight bill, white supercilium and short, yellow legs. ADULT BREEDING: Dark-grey above, heavily barred below. ADULT NON-BREEDING/JUVENILE: Underparts unbarred, the pale grey breast contrasting with the white belly.

VOICE: A clear 8–10-note trill, all on the same pitch.

5 Solitary Sandpiper *Tringa solitaria*

Playero Solitario

Length: 18–21 cm

Occasional migrant; mostly recorded around small pools in the highlands from December to February.

IDENTIFICATION: Body shape and head pattern recall Lesser Yellowlegs, but smaller, with much darker upperparts and shorter, olive legs. Also has thicker bill and white eye-ring. ADULT BREEDING: Dark brown upperparts, finely spotted white. Head, neck and breast streaked blackish. ADULT NON-BREEDING: Plain, dark grey upperparts and breast, less spotted than breeding adult. JUVENILE: As non-breeding adult but upperparts spotted buff.

VOICE: A high "*pee–weet*" or "*pee–wee–weet*".

PLATE 23

❶ Least Sandpiper *Calidris minutilla*

Playero Enan

Length: 13–15 cm

Regular migrant, found in small numbers throughout the year, especially from October to April, on the shoreline or beside pools.

IDENTIFICATION: A tiny wader, the smallest of the *Calidris* sandpipers, and most readily distinguished from the others in all plumages by its yellowish (not black) legs, thin and slightly downturned bill, and frequently 'hunched' posture. The upperparts are also generally darker than the other species.

VOICE: A high-pitched, drawn-out "*kreee*".

❷ Semipalmated Sandpiper *Calidris pusilla*

Playero Semipalmeade

Length: 13–15 cm

Vagrant, recorded occasionally during the northern winter, and occurring in the shore zone.

IDENTIFICATION: Slightly larger than Least Sandpiper, with straighter, thicker bill, black legs and generally paler coloration. Very similar to Western Sandpiper, but bill shorter and blunter, lacking 'drooping' tip. ADULT BREEDING: Upperparts brownish-grey, with very little rufous on head and mantle. Scapulars dark-centred with pale grey edges. Breast and flanks finely streaked dark; belly and under tail white. ADULT NON-BREEDING/FIRST-WINTER: Plain grey-brown above with darker crown and eye stripe. Underparts white with grey wash to sides of breast. JUVENILE: Upperparts more contrasting than in adult, with rufous tones to crown and mantle; fairly prominent white supercilium.

VOICE: A low, short "*churck*".

❸ Western Sandpiper *Calidris mauri*

Playero Occipital

Length: 14–17 cm

Vagrant; recorded most years during the northern winter, in small numbers, and occurring in the shore zone.

IDENTIFICATION: Similar in size and posture to Semipalmated Sandpiper, but bill longer, finer and slightly downcurved at the tip. In breeding and juvenile plumage shows more rufous on head and mantle. ADULT BREEDING: Generally more contrasting than Semipalmated, with bright rufous tones to crown, ear-coverts and scapulars. Neck, breast and flanks heavily streaked black, often extending to belly and undertail. ADULT NON-BREEDING/FIRST-WINTER: Plumage virtually indistinguishable from Semipalmated, though usually the cold grey tones predominate. JUVENILE: Brighter than Semipalmated, showing marked contrast between rufous upper scapulars and grey lower scapulars; crown buff-grey, streaked brown; breast has pale orange-buff wash.

VOICE: High-pitched, thin "*jeet*", very different from call of Semipalmated.

❹ White-rumped Sandpiper *Calidris fuscicollis*

Playero de Rabadilla Blanca

Length: 15–18 cm

Vagrant; only a handful of records, from Fernandina and Floreana.

IDENTIFICATION: Similar in size and structure to Baird's Sandpiper, with long wings extending well beyond the tail when at rest. Greyer overall than Baird's with slightly more drooping black bill, usually showing a pale base to the lower mandible. In flight shows prominent white uppertail-coverts. ADULT NON-BREEDING/FIRST-WINTER: Upperparts grey-brown, lacking buff tones of Baird's, but showing similar pale fringes to the mantle and wing-coverts. Breast lightly streaked grey; rest of underparts white. JUVENILE: Usually more brightly coloured than Baird's, with rich chestnut tones to scapulars and dark-streaked crown, bordered by prominent white supercilium; breast and flanks streaked black under a grey wash.

VOICE: Flight call differs from Baird's: a thin, high-pitched "*cheeet*".

❺ Baird's Sandpiper *Calidris bairdii*

Playero de Baird

Length: 14–17 cm

Vagrant, seen very occasionally on migration between Arctic breeding grounds and South American wintering areas, occurring in the shore zone.

IDENTIFICATION: A small, fairly short-billed wader lacking distinctive plumage characteristics, but its long wings give it an attenuated shape on the ground, similar to that of White-rumped Sandpiper. ADULT NON-BREEDING/FIRST-WINTER: Upperparts and breast-band greyish-brown with buff tones to head and breast. Mantle and wing-coverts show pale fringes, giving a slightly scaly appearance. JUVENILE: Similar to non-breeding adult but upperparts more strongly scaly.

VOICE: A low, trilling "*preeet*".

PLATE 2

1 Red (or Grey) Phalarope *Phalaropus fulicaria*

Falaropo Roj

Length: 20–22 cm
Wingspan: 37–40 cm

Regular migrant, recorded regularly during the northern autumn and winte
mainly at sea. Almost certainly under-recorded due to its similarity to th
much commoner Red-necked Phalarope.

IDENTIFICATION: Slightly larger and more bulky than Red-necked Phalarope, with shorte
and thicker bill. ADULT FEMALE BREEDING: Black crown, white face and brick-red neck an
underparts; mantle black with bright buff fringes. ADULT MALE BREEDING: Similar to breedin
female, but duller and smaller. ADULT NON-BREEDING/FIRST-WINTER: Totally different fron
breeding plumage: pale grey above and all-white below, with black hindcrown and eye-patch
JUVENILE: Upperparts similar to breeding male but neck, breast and flanks buffy-pink an
belly white. Usually shows a dark smudge behind the eye.

VOICE: Call is a short, sharp "*pik*".

BEHAVIOUR: Feeds in similar fashion to Red-necked Phalarope, frequently spinning t
obtain prey.

2 Red-necked Phalarope *Phalaropus lobatus*

Falaropo Norteñc

Length: 18–19 cm
Wingspan: 31–34 cm

Common migrant, particularly between August and April, with large flock
often occurring between the southern and central islands from Decembe
to January. Found at sea or on saltwater lagoons.

IDENTIFICATION: The smallest and daintiest phalarope; thin-necked and very thin-billed
As with all phalaropes, the male is less brightly coloured than the female in breeding plumage
reflecting their reversed roles in incubation and rearing young. ADULT FEMALE BREEDING
Bright chestnut-red sides to neck; slate-grey crown, face, sides of neck and breast; white
throat and eye-spot; dark mantle, with bright buff lateral stripes. ADULT MALE BREEDING
Similar to breeding female but duller. ADULT NON-BREEDING/FIRST-WINTER: Predominantly
grey above and white below, with black crown and eye-patch (usually slightly wider than in
Red Phalarope, which is very similar in this plumage). Upperparts more contrasting than in
Red Phalarope, with whitish fringes forming stripes along edge of mantle and scapulars. In
flight shows darker central uppertail-coverts than Red. JUVENILE: Resembles non-breeding
adult, but dark brown above and buffy-pink below.

VOICE: Call is a shrill "*kip*", reminiscent of Sanderling.

BEHAVIOUR: Unlike other waders, phalaropes are normally seen swimming rather than on
land. Food is picked from the water surface, disturbed by the bird's rapid spinning motion.
Highly gregarious and often extremely tame.

3 Wilson's Phalarope *Phalaropus tricolor*

Falaropo de Wilson

Length: 20–22 cm
Wingspan: 37–40 cm

Regular migrant; mainly from August to December. Unlike the other
phalaropes, this species is usually found on coastal lagoons and occasionally
inland lakes.

IDENTIFICATION: The largest phalarope, with long legs, and a long, straight and very
thin bill. Frequently seen on land, where it resembles a pot-bellied Lesser Yellowlegs. ADULT
FEMALE BREEDING: Unmistakable, with bold black band extending from bill through eye and
down sides of neck, becoming reddish-chestnut; upperparts grey with chestnut bands. ADULT
MALE BREEDING: Similar to breeding female but chestnut plumage replaced by dark brown.
ADULT NON-BREEDING/FIRST-WINTER: Predominantly pale grey above and white below; legs
are bright yellow (black in breeding plumage). JUVENILE: Resembles breeding male, but without
brown neck-band; breast buff.

VOICE: Usually silent, but occasionally gives a soft croaking note in flight.

BEHAVIOUR: Rarely reported at sea. Like the other phalaropes, gregarious and confiding.

PLATE 25

E **1** Galápagos Hawk   *Buteo galapagoensis*   Gavilán de Galápago

Length: *c.*55 cm
Wingspan: *c.*120 cm

Resident and sedentary. Found on most of the islands and occurring in a habitats from the coast to the highlands. Only breeds in drier areas in th lowlands, nesting throughout the year with a peak in activity in June an July. Population much reduced in recent years and now only about 13 breeding territories known. Conservation Status: VULNERABLE.

IDENTIFICATION: The only regularly occurring diurnal raptor readily identified by broad wings and dark plumage. Sexes alike although females appreciably larger than males. ADULT: Plumag dark sooty-brown; tail grey with inconspicuous blackish bands; leg and cere yellow. Flight feathers noticeably paler than rest of wing ir flight. JUVENILE: Blackish-brown above, heavily mottled with buf underparts pale buff, spotted and streaked brown; tail pale buff wit narrow black bands.

VOICE: Gives a range of calls, but typically a far-carrying "*kee–kee–kee...*".

BEHAVIOUR: Often seen soaring on updraughts and occasionally hovers. Feeds both by predation and scavenging. Females mate with more than one male, all birds helping to rea the young.

2 Osprey   *Pandion haliaetus*   Aquila Pescadora

Length: *c.* 60 cm
Wingspan: *c.* 150 cm

Vagrant; recorded annually in small numbers, mainly between June and January.

IDENTIFICATION: A large, long-winged raptor with unmistakable combination of white underparts and brown upperparts; conspicuous black line through eye contrasts with white throat and pale crown; legs grey. In flight, underwings pale with obvious dark mark in carpal area; wings are held noticeably angled and often rather arched, reminiscent of a large gull.

BEHAVIOUR: Feeds exclusively on fish which are caught by plunge-diving, feet-first.

3 Peregrine   *Falco peregrinus*   Alcón Peregrino

Length: 39–50 cm
Wingspan: 95–115 cm

Vagrant; recorded annually in small numbers, mainly between November and March.

IDENTIFICATION: The only falcon recorded; a powerful flier, with pointed, rather broad-based wings and slow, stiff wingbeats, often appearing quite stocky. Sexes alike, although females appreciably larger than males. ADULT: Upperparts blue-grey; underparts white, narrowly barred grey; crown black, extending to prominent moustachials which contrast with white throat; legs, cere and eye-ring yellow. JUVENILE/IMMATURE: Similar to adult but browner overall with streaked rather than barred breast.

BEHAVIOUR: Feeds principally on birds which are pursued at speed and caught in flight, often following a stoop from a great height.

PLATE 2

❶ Common Nighthawk *Chordeiles minor* Aguaitacamino Migratori

Length: 24 cm
Wingspan: 38 cm

Scarce migrant; recorded occasionally in small numbers.

IDENTIFICATION: The only nightjar recorded, appearing rathe long-winged and falcon-like in flight. In flight, prominent oval-shaped white patch in primarie is diagnostic. Upperparts generally brownish-grey with buff and white streaks and spot underparts finely barred blackish-brown and whitish-buff. Throat patch white, particular noticeable in male, although sexes otherwise alike.

BEHAVIOUR: Mainly nocturnal, feeding in flight; most likely to be seen at dawn and dusk During the day rests on the ground and occasionally on posts.

e ❷ Barn Owl *Tyto alba* Lechuza Blanca; Lechuza de Campanario

Length: 26 cm
Wingspan: 68 cm

Uncommon resident; endemic subspecies *punctatissima*. Population estimate at *c.* 9,000 birds. Most numerous in the lowlands, where breeding has bee recorded throughout the year. Also occurs in the highlands where breedin takes place mainly from November to May. Nests in cavities, mainly in o close to the ground but sometimes in holes in trees.

IDENTIFICATION: A medium-sized owl identified b combination of greyish golden-brown upperparts with small white spots, dusky underparts with faint black spotting, heart-shaped facia disc and dark eyes. In flight, underwings appear white.

VOICE: A rather high-pitched, hissing "*shrreeee*", although not ofter heard.

BEHAVIOUR: Mostly nocturnal, although sometimes seen huntin on the ground or from a perch during early morning and late evening flies with slow, deep wingbeats and occasional glides. Sometimes found roosting during the day in buildings.

e ❸ Short-eared Owl *Asio flammeus* Lechuza de Campo

Length: 34–42 cm
Wingspan: 90–105 cm

Uncommon resident; endemic subspecies *galapagoensis*. Population estimated at *c.* 9,000 birds. Found throughout the islands in a range of habitats, breeding from November to May and nesting mainly on the ground.

IDENTIFICATION: A fairly large, rather long-winged, brown owl with short, often inconspicuous, ear-tufts. Upperparts heavily mottled and streaked dark brown and buff; wings dark brown with buff spots; underparts slightly paler than upperparts with brown streaking, heaviest on the breast; facial disc dusky brown, bordered with narrow black and white lines; eyes yellow. In flight, underwings pale with black crescent in carpal area and black wing-tips. Sexes alike, although females appreciably larger than males.

VOICE: Generally silent, but during the breeding season gives a barking "*chef–chef–chef...*" alarm call, and, when displaying a "*whoo–whoo–whoo...*".

BEHAVIOUR: Most active during the early morning and late evening, quartering the ground with slow, deep wingbeats, somewhat reminiscent of a harrier. Tends to feed nocturnally in areas where Galápagos Hawk is present.

PLATE 27

1 Belted Kingfisher *Megaceryle alcyon* Martín Pescador Migratori

Length: 28–33 cm

Annual migrant visitor, in small numbers, mainly between October an March. Breeds in North America, wintering chiefly in Central America.

IDENTIFICATION: Unmistakable; the only kingfisher recorded in Galápagos. Stockil built, with large, crested head and strong, dagger-like bill. Adults are blue-grey above, with broad white collar. Underparts mainly white with, in male, a single blue-grey breast-ban and in female an additional rufous band across the upper belly, extending to the flank Juveniles resemble adults, but have rufous tones to the breast-band.

VOICE: A very noisy species, with a harsh rattle "*kek–kek–kek–kek*".

BEHAVIOUR: Perches or hovers over water, plunging head first to catch fish.

2 Eared Dove *Zenaida auriculata* Paloma Sabaner

Length: 22–28 cm

Vagrant; recorded only once. Occurs throughout South America.

IDENTIFICATION: Larger than Galápagos Dove, with mor uniform coloration. Upperparts olive-brown, with black spots on wings. Head and underpart pinkish-buff, with pale grey crown and black streaks behind and below the eye. Tail pointed with outer feathers edged white and grey. Legs red.

3 Feral Pigeon *Columba livia* Paloma Bravía

Length: 31–34 cm

A fairly common, introduced resident occurring around human habitation

IDENTIFICATION: The familiar pigeon of cities worldwide, whose ancestor is the wild Rock Dove. Occurs in many colour variations, some birds retaining the predominantly grey plumage of the Rock Dove, with double black wing-bar and purple-green iridescent neck and upper-breast, others having brown, black or white elements in numerous permutations.

E 4 Galápagos Dove *Zenaida galapagoensis* Paloma de Galápagos

Length: 18–23 cm

A generally common resident, though less so on Floreana, San Cristóbal and Santa Cruz. Found chiefly in the arid zone, where breeds from February to June, soon after the onset of rains, nesting on the ground, beneath lava or cacti. Two subspecies are recognised: *galapagoensis* (all islands except Darwin and Wolf) and *exsul* (Darwin and Wolf).

IDENTIFICATION: Unmistakable; the only dove likely to be encountered away from human habitation. ADULT: Head, neck and breast reddish-brown, belly buffish. Distinctive pattern on ear-coverts: a white stripe bordered by black lines. Iridescent bronze-green patch on side of neck. Scapulars and wing-coverts black-spotted with white edging. Eye-ring bright blue; legs red. JUVENILE: Similar to adult but plumage generally duller.

VOICE: Displaying males have a very deep, soft, cooing call.

BEHAVIOUR: Often very confiding.

PLATE 28

1 Dark-billed Cuckoo *Coccyzus melacoryphus* Cuclill

Length: 27 cm

Resident, fairly common on Isabela, Fernandina, Santa Cruz, San Cristób and Floreana but less common on Santiago and Pinzón.

IDENTIFICATION: A slim cuckoo, with a long tail and black slightly decurved bill. ADULT: Greyish-brown above, buff below. Hea dark grey, with blackish mask behind eye. Tail bronze above, blac with white tips below. JUVENILE: Similar to adult but generally dulle with buff tips to wing-coverts, and no white tips to underside (tail.

VOICE: A low chuckle, "*cu–cu–cu–cu–culp–culp–culop*".

BEHAVIOUR: A secretive species, more often heard than seen, an usually glimpsed disappearing into undergrowth.

2 Black-billed Cuckoo *Coccyzus erythrophalmus* Cuclillo Pico Negr

Length: 30 cm

Vagrant; a single record of a dead immature in 1970. Breeds in eastern Nort America and winters in north-west South America.

IDENTIFICATION: Slightly larger than Dark-billed Cuckoo, though similar in structure ADULT: Upperparts brown, underparts white. Head the same colour as the mantle, lacking Dark-billed's mask, bill black; prominent red eye-ring. Undertail grey with narrow whit tips. JUVENILE: Underparts tinged buff, but always whiter than Dark-billed. Eye-ring pal yellow. Undertail paler than that of adult.

✓ 3 Smooth-billed Ani *Crotophaga ani* Garrapatero Común

Length: 35 cm

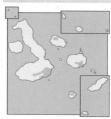

Fairly common introduced resident, occurring in the highlands of Isabela Santa Cruz, Floreana and Santiago.

IDENTIFICATION: Unmistakable, apart from close similarity to Groove-billed Ani (differences from which are described under that species). The only large, entirely black landbird found in Galápagos. ADULT: Plumage glossy-black. Wings broad and relatively short; tail very long, broadening towards the end. Bill very big, black and smooth, flattened laterally with a high arch to the upper mandible, curving down to the forehead. JUVENILE: Similar to adult but plumage generally browner and bill smaller.

VOICE: Has several calls, including a whining, ascending whistle and a harsh two-note alarm, repeated several times.

BEHAVIOUR: Gregarious, seen in small flocks, often in the vicinity of cattle where they feed on insects which they flush.

4 Groove-billed Ani *Crotophaga sulcirostris* Garrapatero Asurcado

Length: 32 cm

The status of this species is uncertain, owing to confusion with the Smooth-billed Ani. A few substantiated records apparently exist.

IDENTIFICATION: In general shape and plumage, indistinguishable from Smooth-billed Ani. However, the bill, although still heavy, is noticeably smaller, and in particular lacks the protruding arch on the upper mandible. At close range, grooves on the upper mandible may be visible, though they are lacking in juveniles.

VOICE: A liquid "*whee–hoo*", accented on the first syllable, a clucking note, and various other calls.

BEHAVIOUR: Similar to that of Smooth-billed Ani.

PLATE 2

❶ Chimney Swift *Chaetura pelagica* Vencejo de Chimer

Length: 13 cm

Vagrant, recorded only once. Breeds in eastern North America, wintering in weste
South America.

IDENTIFICATION: A small, dark swift with curved wings, narrowing slightly as they join the cig
shaped body, and short, square tail. Its flight is fast and direct, very different from that of the hirundin

Ⓔ ❷ Galápagos Martin *Progne modesta* Golondrina de Galápag

Length: 15 cm

Uncommon resident. Present on the central and southern islands of the archipela
in small numbers, most numerous on Isabela, occurring chiefly in the highlan
Probably breeds during the warm/wet season, nests and breeding adults having be
recorded in March.

IDENTIFICATION: The only all-dark hirundine apart from t
uncommon Purple Martin. ADULT MALE: Dark blue overall, with broa
pointed wings and slightly forked tail. ADULT FEMALE/JUVENILE: Soot
black above and dark brown below, usually appearing all-dark in fligh

VOICE: A short, warbling song, a twittering flight call "*cher–cher*", and
high-pitched alarm call are the most frequently heard vocalisations.

BEHAVIOUR: Flight action consists of alternating flaps and glides.

❸ Purple Martin *Progne subis* Golondrina de Iglesi

Length: 19 cm

Infrequent visitor in small numbers, most often on Española and Santa Cruz. Possib
overlooked due to similarity to Galápagos Martin.

IDENTIFICATION: Larger and bulkier than Galápagos Martin with similar flight action. ADUI
MALE: Plumage closely resembles that species, but has small white patch of feathers on sides of low
back, although this is usually concealed. ADULT FEMALE AND JUVENILE: Duller than male, with sooty
grey upperparts and contrasting pale grey underparts, and faint streaking on the breast and flank
Distinguished from Galápagos Martin by paler breast.

BEHAVIOUR: Flight similar to that of Galápagos Martin.

❹ Barn Swallow *Hirundo rustica* Golondrina de Horquill

Length: 18 cm

Uncommon migrant, occurring mostly during the northern winter. Breed
throughout North America, migrating to South America.

IDENTIFICATION: Typical hirundine with long, deeply forked tail in adult. ADULT: Head an
upperparts dark blue, forehead and throat chestnut, underparts cinnamon, with blue breast-band. I
flight shows white spots on spread tail. JUVENILE: Duller and browner than adult; tail lacks long streamer
but still noticeably forked.

BEHAVIOUR: Has characteristic elegant, swooping flight, often low over the ground or water.

❺ Bank Swallow (or Sand Martin) *Riparia riparia* Golondrina Parda

Length: 12 cm

Vagrant or scarce migrant, reported chiefly from the eastern islands. Breeds widel
throughout North America, wintering in South America.

IDENTIFICATION: A small brown and white hirundine with a short, square tail, unlike any other
member of the family found in Galápagos. Upperparts mouse-brown, underparts white except for a
brown breast-band, narrowest in the middle. Tail very slightly forked, but looks square when spread.

BEHAVIOUR: Typical graceful and rapid flight of a hirundine, but more direct than that of Barn
Swallow. Usually feeds over water.

❻ Cliff Swallow *Petrochelidon pyrrhonota* Golondrina Risquera

Length: 13 cm

Vagrant; only a few records.

IDENTIFICATION: Shorter than Barn Swallow, largely because square
tail lacks streamers. In flight the pale rump also helps to distinguish it from this species. ADULT: Upperparts
blue-black, with paler hindneck. Buff forehead contrasts with dark blue crown; sides of head and throat
chestnut except for dark central throat patch. Underparts greyish-white. JUVENILE: duller than adult,
with pale throat and dark forehead.

BEHAVIOUR: Similar to Barn Swallow, though has more soaring flight when feeding.

PLATE 3

e **1** Vermilion Flycatcher *Pyrocephalus rubinus* Pajaro Bru

Length: 14 cm

Common resident, breeding on all the main islands, mostly in the highland Two endemic subspecies occur: *nanus* which is found on all the larger islan except for San Cristóbal; and *dubius* which is confined to San Cristóbal. T two forms are sometimes treated as endemic species and referred to Galápagos Vermilion Flycatcher and San Cristóbal Vermilion Flycatch respectively. Breeds during the warm/wet season from December to Ma nesting in trees and bushes.

IDENTIFICATION: Smaller and daintier than Large-bille Flycatcher. ADULT MALE: Unmistakable, with brilliant red crown an underparts, black mask and upperparts. ADULT FEMALE AND JUVENIL Brown above and yellow below, with a whitish supercilium and pa throat and chin; lacks wing-bars.

The subspecies *dubius* differs from *nanus* in being slightly smaller and generally paler in a plumages

VOICE: During display flight, male has tinkling "*pitty–see, pitty–see*" song, interspersed wit bill-snapping.

BEHAVIOUR: Typical flycatcher, chasing insects in the air and foraging on the ground.

2 Eastern Kingbird *Tyrannus tyrannus* Pitirre American

Length: 22 cm

Vagrant; a single record in 1983.

IDENTIFICATION: A stocky flycatcher, with an upright postur when perched. Adult has a black head, slate-grey upperparts, white underparts washed pal grey across the breast, and a black tail with a distinctive white terminal band. Juvenile i browner above, including the head, and has a darker breast.

E **3** Large-billed (or Galápagos) Flycatcher *Myiarchus magnirostris*

Length: 16 cm

Papamoscas

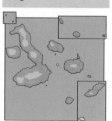

Common resident, occurring on all the islands except Darwin, Wolf and Genovesa. Commonest in the lowlands, though also found in the highlands. Breeds during the warm/wet season, from December to May, nesting in trees and bushes.

IDENTIFICATION: Head and upperparts brownish-grey; chin, throat and breast grey; belly pale yellow, less bright than that of female Vermilion Flycatcher. Closed wing shows a faint double wing-bar. Bill quite thick and black.

VOICE: A liquid "*wheet–wheet–wheet*" call and a melodious song.

BEHAVIOUR: Extremely tame and inquisitive.

PLATE 3

① Blackpoll Warbler *Dendroica striata*

Reinita Rayad[...]

Length: 14 cm — Vagrant; a single record in 1976.

IDENTIFICATION: ADULT MALE: In breeding plumage has blac[...] cap, white cheeks, bold black streaks on olive-grey upperparts, white underparts and tw[...] white wing-bars. FEMALE: Upperparts olive-grey, including crown; dingy white underpart[...] with fainter streaking than male; and distinct double wing-bars. JUVENILE: Similar to femal[...] but throat and breast yellowish.

② Bananaquit *Coereba flaveola*

Reinita Comú[...]

Length: 15 cm — Vagrant; one record.

IDENTIFICATION: A small, active bird with a sharp-pointed[...] decurved bill and rather short tail. Unlikely to be confused with any other species in Galápagos[...] ADULT: Upperparts and tail dark olive with a pale yellow rump; crown and eye-stripe black[...] separated by a long, white supercilium; throat and upper breast grey; rest of underparts lemo[...] yellow. JUVENILE: Duller than adult, with less conspicuous supercilium.

BEHAVIOUR: Feeds energetically on fruits and nectar; often very tame.

③ Red-eyed Vireo *Vireo olivaceus*

Julián Chiuí Ojirajo[...]

Length: 11 cm — Vagrant; few records.

IDENTIFICATION: Recalls a large, thick-billed warbler, with[...] distinctive head pattern. ADULT: Mantle and tail dark olive-brown, underparts pale buffy-white; crown blue-grey, bordered with black line; supercilium white, contrasting with black[...] eye-stripe. The red eye is only visible at close range. JUVENILE: Similar to adult but with yellow[...] wash to underparts and brown iris.

④ Yellow Warbler *Dendroica petechia*

Canario María[...]

Length: 12 cm — Common resident; near-endemic race *aureola* (also occurs on the Cocos[...] Islands). Breeds on all the islands and found in all habitats, nesting during[...] the warm/wet season from December to May.

IDENTIFICATION: Unmistakable; the only bright yellow passerine in Galápagos. ADULT MALE: Upperparts olive-green, with some yellow edgings on darker wings and tail; face and underparts golden-yellow, with faint chestnut streaking on breast and flanks. Crown has reddish-brown patch, variable in size. ADULT FEMALE: Lacks the crown patch of male, and has olive head and upperparts, grey breast and pale yellow belly. JUVENILE: Generally greyer than female but with traces of yellow in plumage.

VOICE: A series of clear, high-pitched "*swee–swee–swee*" notes, followed by a brief staccato warbling.

BEHAVIOUR: Feeds on insects, caught by hawking like a flycatcher or by picking them off the ground.

PLATE 3

1 Cedar Waxwing *Bombycilla cedrorum*

Miracielit

Length: 18 cm

Vagrant; very few records.

IDENTIFICATION: Unmistakable due to combination of black mask and prominent brown crest. Plumage sleek brown overall, darker on the wings and shading to yellow on belly and white on undertail-coverts. Tail has broad yellow tip and secondaries have red wax-like tips, usually concealed.

VOICE: A thin, high-pitched "*tseee*".

2 Summer Tanager *Piranga rubra*

Cardenal Migratori

Length: 20 cm

Vagrant; very few authenticated records. Other tanagers have been recorded but were not specifically identified; these are most likely to have been th species which breeds in North America and winters as far south as north-we: South America.

IDENTIFICATION: ADULT MALE: Unmistakable; bright red overall, with a stout pale bi and beady black eye. ADULT FEMALE/IMMATURE: Olive-grey above, orange-yellow below.

VOICE: Call is a sharp, rattling "*chi–pi–tuck*", with the emphasis on the last syllable.

3 Rose-breasted Grosbeak *Pheuticus ludovicianus*

Picogordio Degollado

Length: 20 cm

Vagrant; very few records.

IDENTIFICATION: A large, chunky finch with a very big triangular bill. ADULT MALE: Head, back and tail black; belly, rump and wing-bars white diagnostic rose-red triangle on the breast. In flight, red underwing is conspicuous. ADULT FEMALE/IMMATURE: Dark brown above, buff below, with heavy, dark streaking on mantle and underparts, narrow white crown-stripes and wing-bars, and, in flight, yellow underwing.

4 Indigo Bunting *Passerina cyanea*

Azulillo

Length: 14 cm

Vagrant; only recorded once.

IDENTIFICATION: A typical bunting in shape, with thick, seed-eating bill and rather broad, slightly forked tail. ADULT MALE: Brilliant turquoise blue overall in full sunlight, but can appear black in poor light or at a distance. ADULT FEMALE: Dull brown above, paler beneath, with faint streaking on the breast. IMMATURE/NON-BREEDING MALE: Generally dark brown with blue tinges.

5 Bobolink *Dolichonyx oryzivorus*

Tordo Arrocero

Length: 18 cm

Regular migrant, mainly from October to December, found chiefly in cultivated areas. Recent records from San Cristóbal indicate that the species may be resident there and could be breeding.

IDENTIFICATION: A large chunky passerine with distinctive pointed tail feathers and a conical, pointed bill. ADULT MALE: Head, underparts, wings and tail black; scapulars and rump white; hindneck buff. FEMALE/ JUVENILE: Buff-brown overall, with darker streaking on back, rump and (female only) on flanks. Head striped dark brown on sides of crown and behind eye.

VOICE: Flight call is repeated "*link*".

PLATE 33

E **1** Chatham (or San Cristóbal) Mockingbird *Nesomimus melanotis*

Cucuve de San Cristóbal

Length: 25 cm

Common resident on San Cristóbal, breeding from October to April and nesting in trees or cacti. Range does not overlap with any other species of mockingbird.

IDENTIFICATION: Similar in size and structure to Galápagos Mockingbird but intermediate in plumage between that species and Hood Mockingbird. Shows a narrow white collar, prominent streaking on flanks and sides of breast, which often extends to form a diffuse breast-band, and a distinct dark malar stripe. Iris greenish.

VOICE: Loud and melodious but variable.

E **2** Charles (or Floreana) Mockingbird *Nesomimus trifasciatus*

Cucuve de Floreana

Length: 28 cm

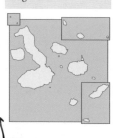

Rare resident, now confined to the islets of Champion-by-Floreana and Gardner-by-Floreana where the population is estimated to be around 150 birds. Conservation status: ENDANGERED. Range does not overlap with any other species of mockingbird. Little information is available on the species' breeding biology, although it is known to nest on Prickly Pear *Opuntia* cacti.

IDENTIFICATION: Similar to Galápagos Mockingbird but bill slightly longer and plumage shows extensive dark smudging on flanks and less streaked upperparts. Lacks the dark ear-coverts and white collar of the other mockingbird species. Iris red-brown.

VOICE: Loud and melodious but variable.

E **3** Galápagos Mockingbird *Nesomimus parvulus*   Cucuve de Galápagos

Length: 25 cm

Locally common resident, breeding from October to April and nesting in trees or cacti. Range does not overlap with any other species of mockingbird. Six subspecies are recognised: *barringtoni* (Santa Fé); *bauri* (Genovesa); *hulli* (Darwin); *parvulus* (Fernandina, Isabela, Santa Cruz, Seymour and Daphne); *personatus* (Pinta, Marchena, Santiago and Rábida) and *wenmani* (Wolf).

IDENTIFICATION: A medium-sized, streaked landbird with long tail, short wings and narrow, moderately long, decurved bill. Upperparts dark grey-brown; wings dark brown with feathers edged and tipped white. Tail dark with white tip. Underparts white, extending up side of neck to form a broad collar; some streaking on flanks and side of breast. Does not generally show a distinct malar stripe. White supercilium contrasts with crown and dark ear-coverts. Iris colour varies from reddish-brown to yellowish-green.

VOICE: Loud and melodious but variable.

E **4** Hood Mockingbird *Nesomimus macdonaldi*   Cucuve de Española

Length: 28 cm

Common resident on Española and Gardner-by-Española, breeding from March to April and nesting in trees or cacti. Range does not overlap with any other species of mockingbird.

IDENTIFICATION: The largest of the mockingbirds in Galápagos with the longest and heaviest bill, and the longest legs. Plumage similar to Galápagos Mockingbird but has a narrower and less conspicuous white collar, and shows heavier streaking on flanks and sides of breast which often extends to form a diffuse, mottled breast-band. Prominent dark malar stripe. Iris hazel.

VOICE: Similar to other species of mockingbird but more strident.

BEHAVIOUR: Travels mainly on foot and rarely flies. Often found in flocks outside the breeding season. Can be extremely tame.

PLATE 34

Identification of Darwin's Finches

The 13 species of Darwin's finch that occur on Galápagos pose by far the greatest identification challenge. Although some of the species are readily identified, it is often extremely difficult to separate others with confidence. This is mainly due to the similarity between some of the species and the variation between different populations of the same species. To confuse matters even more, hybrids also occur. It is therefore often impossible to identify every bird seen and confusing individuals are best left unnamed. However, the distribution of the species is reasonably well understood and, although not strictly scientific, it helps in the identification process to be aware that not all the species occur on all the islands! The checklist of species on pages 150–154 includes a summary of the distribution of the finches.

The species fall into three genera: *Geospiza* (ground finches – 4 species, and cactus finches – 2 species); *Camarhynchus* (Vegetarian Finch, tree finches – 3 species, and Woodpecker and Mangrove Finches); and *Certhidea* (Warbler Finch). However, in terms of plumage features the species are best divided into three groups:

- ground finches and cactus finches (adult males are entirely black);
- Vegetarian Finch and tree finches (adult males have black heads and olive or brown streaked bodies); and
- Woodpecker, Mangrove and Warbler Finch (males and females are alike, being mainly brown or grey).

Although there are differences in plumage and structure between these groups and, in some cases, between the species within each group, the key to identification is the shape, structure and relative size of the bill. Photographs of all of the species appear on Plates 35–38, and a description of the main identification features is included in the accompanying text. However, it is not possible to compare directly the bills of each of the species on these plates and the aim of this section is to highlight the key differences between the species in a form that allows direct comparison. The illustrations opposite are based on measured drawings of a typical bill for each species (indicated by a yellow line) overlaid on a photograph.

Ground finches (4 species) – PLATE 35 and cactus finches (2 species) – PLATE 36

PLUMAGE: ADULT MALES: Black with white-tipped undertail-coverts (except in Sharp-beaked Ground Finch where the undertail-coverts are rufous); adult plumage acquired gradually through a sequence of five moults, the black feathering appearing first on the back and subsequently on the throat, breast and belly. Bill pale pinkish or yellowish, becoming black during the breeding season. FEMALES/IMMATURES: Brown with variable amounts of streaking; generally palest on the underparts.

BILL SIZE, SHAPE AND STRUCTURE: Distinguished as a group from the other types of finch by their rather conical bills with a relatively straight bottom edge to the lower mandible. The bills of cactus finches are similar but are more elongated and pointed and generally appear slightly decurved.

Vegetarian Finch (1 species) and tree finches (3 species)– PLATE 37

PLUMAGE: ADULT MALES: Black head (and neck, breast and back in Vegetarian Finch) when mature, the remainder of the plumage being olive-brown with some dark streaking (the belly of Vegetarian Finch is white). Bill pale pinkish or yellowish, becoming black during the breeding season. FEMALES/IMMATURES: Olive-brown with faint streaking; less heavily streaked on the underparts than ground or cactus finches.

BILL SIZE, SHAPE AND STRUCTURE: Distinguished as a group from the other types of finch as both the culmen and the bottom edge of the lower mandible are curved, giving the bill a rather parrot-like appearance.

Woodpecker, Mangrove and Warbler Finches (3 species) – PLATE 38

PLUMAGE: ADULT MALES AND FEMALES/IMMATURES: Sexes alike. Uniform and virtually unstreaked olive, greyish, buff or brown. Bill pale pinkish or yellowish, becoming black during the breeding season.

BILL SIZE, SHAPE AND STRUCTURE: Woodpecker and Mangrove Finches are distinguished from the other types of finch by their rather stout, long and pointed bills with curved culmen. Warbler Finch is distinguished by its small, thin, pointed bill which is very similar to that of a warbler.

deeper at base than length of upper mandible; base of upper mandible typically starts on forehead and forms raised ridge. Culmen sharply curved but bottom edge to the lower mandible relatively straight.

LARGE GROUND FINCH

Very variable but length of upper mandible always greater than depth of bill at base, and will never as dainty or pointed as that of Small Ground Finch. Culmen curved but bottom edge to the lower mandible relatively straight.

MEDIUM GROUND FINCH

Dainty and rather pointed, the culmen being slightly curved and bottom edge to the lower mandible being relatively straight.

SMALL GROUND FINCH

Small and distinctly pointed, proportionately longer and thinner than any of the other ground finches with a straighter culmen and bottom edge to the lower mandible.

SHARP-BEAKED GROUND FINCH

Rather stout, but long and pointed with a distinctly curved culmen – similar to that of a tanager.

WOODPECKER FINCH

Similar to Woodpecker Finch but shorter and less heavy.

MANGROVE FINCH

Small, thin, pointed and slightly decurved, very similar to that of a warbler.

WARBLER FINCH

Subspecies *conirostris* (from Española): heavy and similar to Large Ground Finch but laterally compressed and more elongated and pointed than that species; the upper mandible is always longer than the depth of the bill at base.

LARGE CACTUS FINCH *conirostris*

Subspecies *propinqua* (from Genovesa): similar to that of *conirostris* but smaller, narrower and more pointed (subspecies *darwini* from Darwin and Wolf is similar).

LARGE CACTUS FINCH *propinqua*

Rather distinctive: relatively thick but distinctly elongated and slightly decurved. The culmen is slightly curved and the bottom edge to the lower mandible is relatively straight.

CACTUS FINCH

Relatively short, deep at the base and broad, with distinctly curved culmen and curved bottom edge to lower mandible.

VEGETARIAN FINCH

Moderately long and stout, and distinctly parrot-like; the culmen is sharply curved, mirrored by an upward curve to the bottom edge of the lower mandible. The tips of the mandibles generally cross, a diagnostic characteristic of this species.

LARGE TREE FINCH

Similar to, but not as parrot-like as, Large Tree Finch, being less stout and with a less sharply curved culmen; the tips of the mandibles do not cross.

MEDIUM TREE FINCH

Relatively small and rather stubby, both the culmen and the bottom edge of the lower mandible being noticeably curved.

SMALL TREE FINCH

E **1** **Large Ground Finch** *Geospiza magnirostris* Pinzón Grande de Tierr

Length: 16 cm

Uncommon resident, least numerous on islands where Medium Grour Finch occurs. Occurs in the arid zone where it is usually found singly; do not feed in flocks. Feeds on the ground less often than the other grour finches. Breeds mainly during the warm/wet season.

IDENTIFICATION: The largest of the ground finches with a massiv bill, the depth at the base being equal to the length of the uppe mandible. The base of the upper mandible often begins on the forehea and forms a raised ridge (SEE PAGE 101). ADULT MALE: Wholly blac with white-tipped undertail-coverts. FEMALE/IMMATURE: Brown wit streaked underparts.

VOICE: A rather slow, low-pitched but distinctive song: "*teu–e–e–e …leur*" or "*teu–woo–whu*".

E **2** **Small Ground Finch** *Geospiza fuliginosa* Pinzón Pequeno de Tierr

Length: 11 cm

Abundant and widespread resident. Most numerous in the coastal, arid an transition zones, breeding during the warm/wet season and moving into th highlands at other times of the year. Often forms a symbiotic relationsh with Giant Tortoises and Marine and Land Iguanas, feeding on skin parasite:

IDENTIFICATION: The smallest, most compact ground finch wit a rather dainty, short, pointed bill, the culmen being slightly curve (SEE PAGE 101). ADULT MALE: Wholly black with white-tippe undertail-coverts. FEMALE/IMMATURE: Brown with streaked underparts

VOICE: A rather weak but rapid disyllabic song: "*twichooo-twichooo*" or "*teur–weee*".

E **3** **Medium Ground Finch** *Geospiza fortis* Pinzón Mediano de Tierra

Length: 12·5 cm

Abundant and widespread resident. Most numerous in the coastal, arid anc transition zones, often forming mixed flocks with Small Ground Finches Breeds mainly during the warm/wet season, often moving to the highlands outside this period.

IDENTIFICATION: Intermediate between Small and Large Ground Finch in size. Whilst the size of the bill is very variable, it is never as dainty or as pointed as Small Ground and, unlike in Large Ground, the length of the upper mandible is always greater than the depth of the bill (SEE PAGE 101). ADULT MALE: Wholly black with white-tipped undertail-coverts. FEMALE/IMMATURE: Brown with streaked underparts.

VOICE: Song similar to but slightly louder than Small Ground, often with trisyllabic notes: "*teur–weee–wee*" or "*tee–er–loo*".

Vampire

E **4** **Sharp-beaked Ground Finch** *Geospiza difficilis* Pinzón Vampiro

Length: 12·5 cm

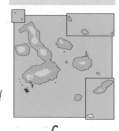

Locally common resident. Three subspecies are recognised: *debilirostris* (Fernandina and Santiago); *difficilis* (Genovesa and Pinta) and *septentrionalis* (Darwin and Wolf). Generally found in the highlands, breeding during the warm/wet season but disperses to the lowlands outside this period; confined to the shore and arid zones on Genovesa, Darwin and Wolf. The population on Wolf and Darwin is often referred to as the Vampire Finch since the birds there regularly feed on blood obtained by pecking at the base of Nazca Booby feathers.

IDENTIFICATION: Very similar to Small Ground Finch but bill longer and more pointed, the culmen being less curved (SEE PAGE 101). ADULT MALE: Wholly black with rufous undertail-coverts. FEMALE/ IMMATURE: Generally darker than the other species of ground finch.

VOICE: A rather feeble song with short notes, reminiscent of Warbler Finch (Plate 38).

PLATE 3

ⓔ ❶ Cactus Finch *Geospiza scandens* Pinzón del Cactu

Length: 14 cm

Locally common resident. Four subspecies are recognised: *abingdoni* (Pinta intermedia* (Santa Fé, Floreana, Santa Cruz, Isabela and Pinzón (although may be extinct on this island)); *rothschildi* (Marchena); and *scandens* (Santiag and Rábida). Restricted to the arid zone and rarely seen away from areas Prickly Pear Cacti *Opuntia*, upon which nests are built. Breeds during th warm/wet season.

IDENTIFICATION: Plumage characteristics similar to the groun finches but bill rather distinctive, being long and pointed an appearing slightly decurved (SEE PAGE 101). ADULT MALE: Wholl black with white-tipped undertail-coverts. FEMALE/IMMATURE: Brow and heavily streaked, particularly around the head and neck whic often appears quite dark, a feature which distinguishes the species from the ground finches

VOICE: Most frequent call is a loud, ringing "*teur–lee, teur–lee*", but also gives a "*teur–teur teur–tweee*". Occasionally sings in flight.

ⓔ ❷ Large Cactus Finch *Geospiza conirostris* Pinzón Grande de Cactu

Length: 15 cm

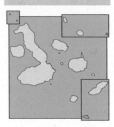

Locally common resident. Three subspecies are recognised: *conirostris* ❷Ⓐ (Española); *darwinii* (Darwin and Wolf); and *propinqua* ❷Ⓑ (Genovesa Does not occur on islands where Cactus Finch is present. Breeds during th warm/wet season. Often feeds on the ground.

IDENTIFICATION: Plumage characteristics similar to the groun finches and Cactus Finch. ADULT MALE: Wholly black with white tipped undertail-coverts. FEMALE/IMMATURE: Dull black or dark gre often with faint white fringes to the feathers on the underparts; th darkest of all the ground and cactus finches.

Subspecies *conirostris* ❷Ⓐ is slightly larger than *propinqua* ❷Ⓑ and *darwinii*, with a heavier bill which is similar to Large Groun Finch but laterally compressed and more elongated and pointed; the upper mandible is alway longer than the depth of the bill at base. Bills of *propinqua* and *darwinii* are smaller, narrowe and more pointed (SEE PAGE 101).

VOICE: A rather distinctive "*chee–yoo–oo*" or "*klee–yoo*".

PLATE 37

E **1 Vegetarian Finch** *Camarhynchus crassirostris* Pinzón Vegetariano

Length: 16 cm

Uncommon resident; generally found in the transition zone but occasionally in the arid and humid zones. Breeds principally during the warm/wet season. Has a distinctly upright stance when perched, feeding quietly and moving rather slowly.

IDENTIFICATION: Distinguished by large size and short, broad bill which is deep at the base (SEE PAGE 101). ADULT MALE: Head, neck, breast and back black when fully mature, the remainder of the plumage being olive-brown and streaked, apart from the belly which is white. Bill pale pinkish or yellowish becoming black during the breeding season. FEMALE/IMMATURE: Upperparts brown and streaked, with unstreaked olive rump. Underparts pale, whitish or yellowish, with dark streaks on breast and flanks.

VOICE: A rather loud and distinctive song which comprises several musical notes followed by a harsh buzz and often ending with a lower-pitched whistle.

E **2 Large Tree Finch** *Camarhynchus psittacula* Pinzón Grande de Árbol

Length: 13 cm

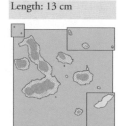

Uncommon resident. Three subspecies are recognised: *affinis* (Isabela, Fernandina, habeli* (Pinta, Marchena); and *psittacula* (Seymour, Santa Fé, Santa Cruz, Floreana, Pinzón, Rábida, Santiago and possibly San Cristóbal). Breeds in the highlands, mainly during the warm/wet season, often moving to lower elevations at other times.

IDENTIFICATION: The largest of the tree finches with a large, rather parrot-like bill, the tips of the mandibles crossing (SEE PAGE 101). ADULT MALE: Head, neck, breast and mantle black when fully mature, the remainder of the upperparts being olive-grey with some dark streaking (however, commonest plumage has black restricted to head and throat). Underparts pale, often with a yellow tinge, but with dark streaks, particularly on upper breast. FEMALE/IMMATURE: Upperparts olive-brown with faint streaking. Underparts paler and virtually unstreaked, sometimes with a yellowish wash.

VOICE: A rather soft, low-pitched song that typically includes a ringing and often repeated "*toorelu*" or "*twee–twee*" and generally ends with a series of high-pitched "*e*" notes.

E **3 Medium Tree Finch** *Camarhynchus pauper* Pinzón Mediano de Árbol

Length: 13 cm

Fairly common resident, restricted to Floreana where most numerous in the highlands, sometimes moving to lower elevations outside the breeding season. Breeds when conditions are suitable, principally during the warm/wet season. Conservation Status: NEAR-THREATENED.

IDENTIFICATION: Similar in size to Large Tree Finch but bill smaller and less parrot-like (the tips of the mandibles do not cross) (SEE PAGE 101). Respective plumages similar to those of Large Tree Finch.

VOICE: Virtually indistinguishable from Large Tree Finch but sometimes gives a five-syllable "*tshoo–tshoo–tshoo–tshoo–tshoo*" and occasionally a "*tsee–tsee–tsee*".

E **4 Small Tree Finch** *Camarhynchus parvulus* Pinzón Pequeño de Árbol

Length: 11 cm

Common and widespread resident. Two subspecies are recognised: *parvulus* (Santiago, Rábida, Santa Cruz, Seymour, Santa Fé, Isabela, Pinzón, Floreana and Fernandina); and *salvini* (San Cristóbal). Most numerous in the highlands but also occurs in the arid zone, particularly outside the breeding season. Breeds principally during the warm/wet season. Often feeds upside-down at the end of branches, reminiscent of a chickadee or tit. Occasionally feeds on the ground.

IDENTIFICATION: The smallest of the tree finches with a small, rather stubby bill (SEE PAGE 101). ADULT MALE: Plumage similar to the other tree finches but underparts generally pale yellow with restricted amount of streaking on upper breast. FEMALE/IMMATURE: Upperparts grey-brown with very faint streaking. Underparts paler and virtually unstreaked. Often has a pale area around eye and a hint of a pale supercilium.

VOICE: Very variable, virtually indistinguishable from other tree finches.

PLATE 38

E **1** **Woodpecker Finch** · *Camarhynchus pallidus* Pinzón Artesan...

Length: 15 cm

Locally common resident. Three subspecies are recognised: *pallidus* (Santiag... Rábida, Pinzón, Santa Cruz and Seymour) **1A**; *productus* (Fernandina an... Isabela) **1B**; and *striatipectus* (San Cristóbal). Most numerous in the highland... although also present and breeds in the lowlands, nesting whenever condition... are suitable. Distinctive behaviour whilst searching for food, often looking an... probing under branches, leaves or rocks in a manner reminiscent of a nuthatch... One of the few species of bird in the world that uses a tool, some individua... having learned to manipulate a twig or thorn in its bill to ease insects out o... crevices.

IDENTIFICATION: Males and females are alike. Upperparts uniform... and virtually unstreaked olive or brown; underparts yellowish or whitish... usually unstreaked but sometimes with fine grey streaking on upper breast. Bill long and rathe... stout with a distinctly curved culmen, similar to that of a tanager (SEE PAGE 101).

VOICE: A rapid, loud and distinctive song which comprises seven or eight notes followed by... prolonged "*seee*". Also occasionally gives other calls, such as a loud "*cheer—keee–e–e–e*" or "*tchurr–tchurr–tchur–e–e–e*".

E **2** **Mangrove Finch** *Camarhynchus heliobates* Pinzón del Mangla...

Length: 14 cm

Very rare resident, confined to dense stands of mangrove on the west coast o... Isabela and possibly the east coast of Fernandina. The population must be ver... small, due to the restricted range of its favoured habitat, and may number n... more than a few dozen birds. Conservation Status: ENDANGERED. Some bird... occasionally use a twig to ease insects out of crevices, using the same techniqu... as the Woodpecker Finch

IDENTIFICATION: Males and females are alike. Similar to... Woodpecker Finch but slightly smaller, with a less heavy bill and... generally greyer plumage which does not show yellowish tinges. The... upperparts are grey-brown with a slight olive hue to the rump, and the... underparts are greyish-white with some grey spotting on breast. Often... shows a distinctive pale area around eye (SEE PAGE 101).

VOICE: A rather loud and distinctive song which comprises two or three syllables, usually repeated... three times: typically "*ds–shed–ee, ds–shed–ee, ds–shed–ee*", occasionally ending with a rising, prolonged "*tseee*".

E **3** **Warbler Finch** *Certhidia olivacea* Pinzón Cantor

Length: 10 cm

Common and widespread resident; found in all vegetation zones, being most... numerous in the humid zone although common in the arid zone on the smaller... islands. Breeds whenever conditions are suitable, principally during the warm/... wet season. Eight subspecies are recognised: *becki* (Darwin and Wolf); *bifasciata*... (Santa Fé); *cinerascens* (Española) **3A**; *fusca* (Marchena and Pinta); *luteola*... (San Cristóbal); *mentalis* (Genovesa) **3B**; *olivacea* (Fernandina, Isabela, Pinzón,... Rábida, Santa Cruz, Santiago and Seymour) **3C**; and *ridgwayi* (Floreana).

IDENTIFICATION: By far the smallest of the Darwin's finches, with... the smallest, narrowest bill which is very similar to that of a warbler.... The plumage of the different subspecies vary: the colour of the upperparts... ranges from pale grey to olive-green, and the underparts from white to... buff. Some subspecies show a distinct pale eye-ring. Whilst males and females generally have... identical plumage, some adult males (particularly of subspecies *olivacea*) develop an orange throat-... patch. The lack of yellow in the plumage distinguishes the Warbler Finch from the slightly larger... Yellow Warbler (see plate 31), which is the only likely confusion species (SEE PAGE 101).

VOICE: A rather feeble but melodious warbling song, usually ending with a high-pitched buzz.

THE REPTILES OF GALÁPAGOS

Introduction

Twenty eight species of reptile have been recorded in Galápagos in recent times. Nineteen of these species are endemic to the archipelago, 11 of which are confined to single islands, and three species have been introduced. There is also possibly an undescribed species of Leaf-toed Gecko on Rábida. The reptiles that occur in Galápagos can be divided into seven 'types'. These are shown in the following table which indicates the number of species recorded and provides a summary of their status.

Type (Plate No.)	Species Recorded	Status			Endemic Species
		Indigenous Residents	Marine Migrants	Introduced	
Tortoise (39)	1	1			1
Turtles (40)	4	1	3		
Iguanas (41)	3	3			3
Lava lizards (42)	7	7			7
Geckos (43)	9 (10?)	6 (7?)		3	5 (6?)
Snakes (44)	3	3			3
Sea snakes (44)	1		1		

The types of reptile

This part of the book provides an introduction to each of the seven 'types' of reptile that occur in Galápagos. It aims to aid initial identification, and the text for each of the groups is cross-referenced to the relevant plate(s) that follow.

Galápagos Tortoise

TORTOISES   **Family: Testudinidae**
1 species recorded   **(Plate 39)**
(indigenous resident), represented by 11 subspecies

Endemic species: GALÁPAGOS (OR GIANT) TORTOISE.

Tortoises are unmistakable reptiles, readily distinguished by their hard, dome-shaped carapace which is formed of bony plates which are fused to the skeleton. They differ from the turtles in being entirely terrestrial, their legs being adapted for walking rather than swimming. The Galápagos Tortoise grows to a very large size (sometimes up to 250 kg) and exhibits two distinct carapace types: dome-shaped and 'saddleback', the latter having a raised ridge at the front edge.

Black Turtle

TURTLES
Family: Cheloniidae
4 species recorded **(Plate 40)**
(1 indigenous resident; 3 marine migrants)

Turtles are unmistakable reptiles. They are almost entirely marine, only coming ashore to lay their eggs. They have hard, rather flattened carapaces which, as with the tortoises, are formed of fused bony plates. They differ from tortoises in having legs which are flattened to form flippers.

Marine Iguana (male)

IGUANAS
Family: Iguanidae
3 species recorded **(Plate 41)**
(all indigenous residents), one represented by 7 subspecies

Endemic species: LAND IGUANA; MARINE IGUANA; SANTA FÉ LAND IGUANA.

Iguanas are medium-sized to largish reptiles with long, tapering tails, rather stout bodies and broad heads. They are distinguished from lizards by their larger size, the presence of a row of spines along the back (particularly prominent in males) and rather broad heads.

LAVA LIZARDS
Family: Iguanidae
7 species recorded **(Plate 42)**
(all indigenous residents)

Endemic species: ESPAÑOLA LAVA LIZARD; FLOREANA LAVA LIZARD; GALÁPAGOS LAVA LIZARD; MARCHENA LAVA LIZARD; PINTA LAVA LIZARD; PINZÓN LAVA LIZARD; SAN CRISTÓBAL LAVA LIZARD.

Lava lizards are small reptiles with long, tapering tails, slim bodies and rather pointed heads. Their toes are long and pointed, quite different from those of the geckos which are the only likely confusion species. The average size of the species varies between islands and the markings are extremely variable even within the same species. Unlike geckos, they are active during the day.

Galápagos Lava Lizard (male)

111

Galápagos Leaf-toed Gecko

GECKOS   Family: Gekkonidae
10 species recorded   (Plate 43)
(7 indigenous residents, one of which may now be extinct; 3 introduced)

Endemic species: BAUR'S LEAF-TOED GECKO; GALÁPAGOS LEAF-TOED GECKO; RÁBIDA LEAF-TOED GECKO (undescribed, possibly EXTINCT); SAN CRISTÓBAL LEAF-TOED GECKO; SANTA FÉ LEAF-TOED GECKO; WENMAN LEAF-TOED GECKO.

Geckos are very small, nocturnally-active reptiles. They resemble lava lizards but have thicker tails and rather broad heads with large eyes. Their toes are diagnostic, having folds of skin which form pads enabling them to climb even smooth vertical surfaces with ease.

Española Snake

SNAKES   Family: Colubridae
3 species recorded (3 terrestrial indigenous residents, two of which   (Plate 44)
are represented by 3 subspecies and the other by 2 subspecies)

Endemic species: FLOREANA SNAKE (includes named subspecies Española Snake and San Cristóbal Snake); GALÁPAGOS SNAKE (includes named subspecies Fernandina Snake and Isabela Snake); SLEVIN'S SNAKE (includes named subspecies Steindachner's Snake).

Snakes are unmistakable reptiles, readily distinguished from all others by their long, thin bodies and the absence of legs.

Yellow-bellied Sea Snake

SEA SNAKES   Family: Hydrophiidae
1 species recorded   (Plate 44)
(migrant marine species)

Similar in structure to the terrestrial snakes but the only species recorded in Galápagos, the Yellow-bellied (or Pelagic) Sea Snake, is entirely marine. It is characterised by its black and yellow pattern and flattened tail, an adaptation for swimming.

The reptile plates

The species illustrated All the species of tortoise, turtle, iguana and snake to have been recorded in Galápagos are illustrated in the following plates. Due to the great similarity between the species of lava lizard and gecko, and since most of the species are found only on single islands where they are the only representative of their group, a few examples only are illustrated. However, all species are covered in the text facing the plates.

The species order For ease of reference, the plates are ordered so that all the potential confusion species are grouped together. The text follows the sequence in which the species appear on the plate, from top to bottom. The species are numbered sequentially and the following annotations are used on the plates:

- **1** Adult (species 1)
- **1m** Male
- **1f** Female
- **1i** Immature

Distribution maps A distribution map accompanies the text for each of the resident species or, in some cases, types of reptile. As with the maps used in the bird section, three 'groups' of islands have been moved to save space (see page 33 for further details). The distribution of the species is shown as solid colour on the maps in this section, as in the 'dummy' map opposite.

English name The English names used in this book for the reptiles are those most commonly cited in recent literature.

Spanish name The Spanish names shown are those which are most regularly used in Galápagos.

Endemic species The names of species which are endemic to Galápagos are preceded by an **E**.

Status The first part of the text for each species provides information on its status. This includes details of whether it is an indigenous resident, a migrant/vagrant or an introduced species, and provides a summary of its breeding season(s) and habitat preferences. The names of endemic subspecies are given in this section where appropriate.

Measurements The maximum recorded length of each species is given.

Identification Details of the features which are key to the identification of the species are given, differences between males and females being described where appropriate. Some species are virtually impossible to identify unless examined in the hand which is not permitted under the Galápagos National Park rules without a special scientific permit.

Behaviour Particular behavioural characteristics which help in the identification of a species are highlighted where appropriate.

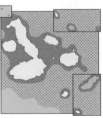

- Breeding range (regular)
- · Breeding site
- Regularly seen at sea
- Occasionally seen at sea
- Rarely seen at sea

PLATE 3

E Galápagos (or Giant) Tortoise *Geochelone elephantopus* Galápag

Length: up to 150 cm over the curve of the carapace, with different subspecies growing to different sizes. Weight: up to 250 kg

Locally common, found mainly in the highlands but often descending lower elevations during the warm/wet season. Endemic to Galápag fourteen subspecies have been described, although only 11 of these st survive: *abindgoni* (Pinta – now EXTINCT in the wild); *becki* (Volcán Wo Isabela); *chathamensis* (San Cristóbal); *darwini* (Santiago); *elephantop* (Cerro Azul, Isabela); *ephippium* (Pinzón); *guntheri* (Sierra Negra, Isabela *hoodensis* (Española) **A**; *microphyes* (Volcán Darwin, Isabela); *porteri* (San Cruz) **I**; and *vandenburghi* (Volcán Alcedo, Isabela). The total populatic is estimated to number *c.*15,000.

IDENTIFICATION: Unmistakable due to its large size and distinctive shape. The Galápag Tortoise exhibits two distinct carapace shapes: dome-shelled and saddle-backed, althoug intermediate forms do occur. The dome-shelled subspecies (*chathamensis, elephantopu guntheri, porteri* **I** and *vandenburghi*) generally have shorter legs and necks than the saddl backed subspecies (*abindgoni, ephippium* and *hoodensis* **A**), in which the front edge of th carapace is raised enabling maximum elevation of the long neck. The carapace in saddleback is usually narrowed at the rear end and these subspecies also have much longer legs tha dome-shelled forms. The carapace of subspecies *microphyes* and *darwini* are intermediate i shape, being rather flattened, although they are not elevated at the front. The carapace shap of subspecies *becki* is very variable, some individuals being dome-shaped, others flattene and others distinctly saddleback-shaped.

In all subspecies the sexes are alike, although males are much larger than females, and have concave plastron (ventral plate) and a noticeably longer and thicker tail.

VOICE: Usually silent but during copulation males emit a loud, deep groan which can b heard over a considerable distance.

BEHAVIOUR: Tortoises are usually only active from about 08.00 until 16.00. Mating ma occur in almost any month of the year but reaches a peak during the warm/wet season, wit eggs usually hatching between December and April. Sexual maturity is attained at the age o about 20–25 years. During the breeding season males chase each other and posture by raisin their heads as high as possible, the dominant individuals being those able to raise their head the highest.

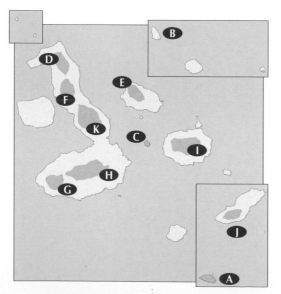

GALÁPAGOS TORTOISE SUBSPECIES

A *hoodensis*

B *abindgoni*

C *ephippium*

D *becki*

E *darwini*

F *microphyes*

G *elephantopus*

H *guntheri*

I *porteri*

J *chathamensis*

K *vandenburghi*

PLATE 4

1 Leatherback Turtle *Dermochelys coriacea*   Tortuga Lar

Length: up to 188 cm   Occasional migrant; only ever seen alive at sea in Galápagos.

IDENTIFICATION: The largest turtle in the world and high distinctive. The carapace is tapered towards the rear and is blac with pale spots covered by leathery skin which forms 7 ridges that run along its entire lengt The head lacks scales and is rounded, although there is a tooth-like projection on the upp mandible.

2 Olive Ridley Turtle *Lepidochelys olivacea*   Tortuga Golfin

Length: up to 76 cm   Occasional migrant; only ever seen alive at sea in Galápagos.

IDENTIFICATION: Carapace greyish-green and rounded, wit 6–9 costals (the outermost large plates) which are unequal in shape. The vertebral scute (central plates) are small and also unequal in shape. The head is quite broad and has a distinctl hooked beak.

3 Hawksbill Turtle *Eretmochelys imbricata*   Tortuga Care

Length: up to 90 cm   Occasional migrant; only ever seen alive at sea in Galápagos.

IDENTIFICATION: Carapace dark brown, often with tortoiseshe markings, and teardrop-shaped, tapering towards the rear, with a slight keel. The vertebra scutes (central plates) overlap and have diagnostic 'W'-shaped rear margins (although thi can sometime be difficult to see). The head is rather pointed with a prominent, hooked beak

4 Black (or Pacific Green) Turtle *Chelonia mydas*   Tortuga Negra

Length: up to 84 cm   Locally common resident; the subspecies that occurs in Galápagos is the eastern Pacific form a*gassizi* which some authorities treat as a distinct species

IDENTIFICATION: Carapace blackish to olive-brown and teardrop-shaped, tapering towards the rear, with a slight keel. The vertebral scutes (central plates) are roughly equal in size and are hexagonal with straight edges; they are about the same size as the costals (the outermost large plates). The head is rounded and the beak is, at most, only slightly hooked. The sexes are alike, although females are distinctly larger than males, and males have a concave plastron.

BEHAVIOUR: Almost entirely marine, coming ashore usually only to lay eggs. Egg-laying takes place from December to June, with a peak in activity during February. Egg-laying usually takes place at night.

PLATE 4

E 1 Marine Iguana *Amblyrhynchus cristatus* Iguana Marir

Length: 60–150 cm, different subspecies growing to different sizes: individuals on Isabela, Santa Cruz and Fernandina are the largest and those on Genovesa and Española the smallest. (Hatchlings are similar in size to lava lizards.)

Abundant in the shore zone on all the islands. Seven subspecies are recognise *albemarlensis* (Isabela) ; *cristatus* (Fernandina) ; *hassi* (Santa Cru ; *mertensi* (San Cristóbal and Santiago); *nanus* (Genovesa); *sielman* (Pinta); and *venustissimus* (Española and Gardner-near-Española) . Th total population is estimated to number *c.* 250,000. Hybrids with Land Iguar have been recorded on South Plaza.

IDENTIFICATION: Large and dark with variable coloration Distinguished from the land iguanas by the flattened, rather squar nose, an adaptation for feeding on marine algae, and by the lateral flattened tail, an adaptation for swimming. ADULT MALE: Row of lor spines on head and along back and tail. Variable pattern on body which ranges from red and blac (*venustissimus* on Española), through green, yellow and black (e.g. *albemarlensis* on Isabela) predominantly dark sooty-grey (*cristatus* on Fernandina). During the mating season assume considerably brighter coloration. ADULT FEMALE/JUVENILE: Much smaller than adult males an wholly dark, with ridge of short spines along back and tail.

BEHAVIOUR: Found predominantly along rocky shores, females or immatures often baskin together in large numbers. The only marine lizard in the world, feeding on marine algae eithe obtained from the splash zone or by diving close to the shore; can spend up to an hour unde water. Sometimes climbs trees or cacti to bask in the sun. Strongly territorial, particularly durin the breeding season. Male shakes head rapidly up and down and exhales loudly when agitated Breeds from November to December. Males take up to eight years to reach sexual maturity.

E 2 Land Iguana *Conolophus subcristatus* Iguana Terrestr

Length: up to *c.*100 cm (hatchlings are similar in size to lava lizards)

Locally fairly common, inhabiting the arid zone. Hybrids with Marine Iguar have been recorded on South Plaza but do not appear to be very long-lived

IDENTIFICATION: Large and pale, generally yellowish-orange i colour with darker brown back. Distinguished from Marine Iguana by the rather pointed nos and from the Santa Fé Land Iguana by the more restricted row of spines along the back. ADUL MALE: Row of long spines along back of neck, with shorter spines on head and along upperback Mature individuals assume some red coloration, particularly during the mating season. ADUL FEMALE: Considerably smaller and less brightly coloured than adult male, with shorter spines.

BEHAVIOUR: Forms small colonies although often found singly. Males are highly territorial defending their territories against intruders by engaging in head-butting battles. Breeding begins i different months on different islands: on Isabela in January, on Fernandina in June, on Santa Cruz in September and on South Plaza in January. Males can take up to 12 years to reach sexual maturity.

E 3 Santa Fé Land Iguana *Conolophus pallidus* Iguana Terrestre Galapagueña

Length: up to *c.*120 cm (hatchlings are similar in size to lava lizards)

Fairly common; confined to Santa Fé, inhabiting the arid zone.

IDENTIFICATION: Large and generally pale, whitish to dark brown in colour, often with large dark brown blotches on back. Distinguished from Marine Iguana by the rather pointed nose and from the Land Iguana by the more extensive row of spines along the back. ADULT MALE: Row of medium-length spines along neck, back and tail. In some individuals the eyes become red. ADULT FEMALE: Considerably smaller than adult male, with shorter spines.

BEHAVIOUR: Found throughout Santa Fé away from the shoreline, forming small colonies but often found singly. Like the other iguanas, males are highly territorial, defending their territories against intruders by engaging in head-butting battles. Mating occurs in February and March. Males can take up to 12 years to reach sexual maturity.

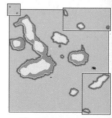

SPECIES

■ **Marine Iguana**

■ **Land Iguana**

■ **Santa Fé Land Iguana**

PLATE 4

E Lava Lizards   *Microlophus (= Tropidurus)* spp.

Lagartija de la

Common in the shore and arid zones. Seven species occur, all of which are endemic. Six of the species
confined to single islands, as reflected by their common names:

1 **Galápagos Lava Lizard** *Microlophus albemarlensis*
(Baltra, Daphne Major, Fernandina, Isabela, North Seymour,
South Plaza, Rábida, Santa Cruz, Santa Fé, and Santiago)

2 **Española Lava Lizard** *Microlophus delanonis*

3 **Floreana Lava Lizard** *Microlophus grayi*
(also occus on Caldwell, Champion, Enderby and
Gardner-near-Floreana)

4 **Marchena Lava Lizard** *Microlophus habellii*

5 **Pinta Lava Lizard** *Microlophus pacificus*

6 **Pinzón Lava Lizard** *Microlophus duncanensis*

7 **San Cristóbal Lava Lizard** *Microlophus bivattatus*

Length: 15–30 cm
Different species grow
to different sizes: the
largest is the
Española Lava Lizard
(up to 30 cm) and the
smallest is the
Floreana Lava Lizard
(rarely more than 15 cm

IDENTIFICATION: Lava lizards are readily identified by their relatively small size an
long, tapering tails, slim bodies and rather pointed heads. The toes are long and pointed an
the skin is scaly, quite different from the geckos which are the only likely confusion specie
Specific identification is impossible without first catching individuals. However, the fact th
no more than one species occurs on any one island does facilitate identification.
ADULT MALE: Very variable in coloration and the pattern of markings on the body. The:
markings are generally adapted to blend in with the substrate upon which they live. Mo
males are dark grey with black speckling, although some are reddish and others have pa
stripes along their sides. When mature they have a black throat and develop a short crest c
spiny scales which extends along the back and which can be raised during display. Males ar
2–3 times heavier than females. ADULT FEMALE: Similar to male in general structure but muc
smaller and lacking the spiny scales along the back. When mature they have a red or orang
throat which in some species (e.g. Española Lava Lizard) extends to the whole head.

BEHAVIOUR: Active during the day. Highly territorial, both males and females defendin
territories against intruders of the same sex. Both sexes challenge intruders initially by performin
'push-ups' with their front legs, although this behaviour is more vigorous in males. Breedin
occurs from December, at the start of the warm/wet season. Males take up to three years t
reach sexual maturity, whereas some females can breed when they are nine months old.

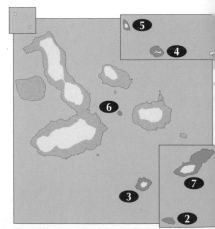

SPECIES

■ Galápagos Lava Lizard

■ Single island endemics

PLATE 4

Geckos *Phyllodactylus* spp. and introduced species

Salamanquesa, Geo

Locally common in the shore and arid zones and around human habitation. Ten species occur in Galápag six of which are endemic (although one may be extinct) and three of which have been introduced recent times.

E **1** **Galápagos Leaf-toed Gecko** *Phyllodactylus galapagoensis*
(Daphne Major, Fernandina, Isabela, Pinzón, Santa Cruz,
and Santiago); three individuals depicted, showing the highly variable coloration of this specie

Length: up to 15 cm

E **2** **Baur's Leaf-toed Gecko** *Phyllodactylus bauri*
(Caldwell, Champion, Enderby, Española, Floreana and Gardner-near-Floreana)

E **3** **Rábida Leaf-toed Gecko** *Phyllodactylus* sp. (undescribed)
(Rábida – possibly Extinct)

E **4** **San Cristóbal Leaf-toed Gecko** *Phyllodactylus leei*
(San Cristóbal)

E **5** **Santa Fé Leaf-toed Gecko** *Phyllodactylus barringtonensis*
(Santa Fé)

6 **Tuberculated Leaf-toed Gecko** *Phyllodactylus tuberculosus*
(San Cristóbal)

E **7** **Wenman Leaf-toed Gecko** *Phyllodactylus gilberti*
(Wolf)

The three introduced species are:

8 *Gonatodes caudiscutatus,* **9** *Lepidodactylus lugubris* and **10** *Phyllodactylus reissi*

IDENTIFICATION: Geckos resemble lava lizards but are readily identified by their small si and rather broad heads with large, dark eyes, the pupil being a vertical slit, and permanent closed, transparent eyelids. Their toes are diagnostic, having folds of skin which form pads enablin them to climb even smooth vertical surfaces with ease. Geckos have soft skin and generally sandy coloured bodies with dark mottling on the upper surface and often a pale line through the ey Specific identification is impossible without first catching individuals. However, only San Cristóba supports more than one of the indigenous species and the introduced species are restricted to area of human habitation, principally on Santa Cruz. Identification can therefore usually be made o the basis of distribution.

BEHAVIOUR: Active only at night when they can be quite vocal, giving high-pitched squeaks During the day they hide under rocks, logs and in cracks. Breed in October and November

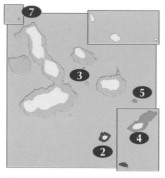

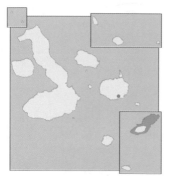

Endemic species

■ **Galápagos Leaf-toed Gecko** **1**

■ **Baur's Leaf-toed Gecko** **2**

■ **Single island endemics** **3** **4** **5** **7**

Non-endemic & introduced species

■ **Tuberculated Leaf-toed Gecko** **6**

■ **Introduced species** **8** **9** **10**

PLATE 4

E **1** Floreana/Española/San Cristóbal Snake *Philodryas biserialis*

Culebra biserialis/hoodensis/eil

Length: up to 120 cm

Locally fairly common. Three named subspecies occur, mainly in the arid a shore zones: Floreana Snake *biserialis* (Champion, Floreana and Gardner-ne Floreana), Española Snake *hoodensis* **1A** (Española and Gardner-near-Españo and San Cristóbal Snake *eibli* (San Cristóbal). Feeds by constricting its pr although slightly venomous.

IDENTIFICATION: Virtually impossible to identify unless examin in the hand, identification being confirmed on the basis of the shap pattern and number of scales on certain parts of the body. Howeve this species only occurs on islands from which the two other speci that occur in Galápagos are absent. As with the other specie predominantly brown with yellow stripes or dark grey with yellow spots on the upperside formin a zigzag pattern.

E **2** Galápagos/Fernandina/Isabela Snake *Alsophis dorsalis*

Culebra dorsalis/occidentalis/helle

Length: up to 120 cm

Locally fairly common. Two named subspecies occur, mainly in the ar and shore zones: Galápagos Snake *dorsalis* (Baltra, Rábida, Santa Cruz, San Fé and Santiago), Fernandina Snake *occidentalis* **2A** (Fernandina an Isabela Snake *helleri* (Isabela and Tortuga). Feeds by constricting its pre although slightly venomous.

IDENTIFICATION: Virtually impossible to identify unless examine in the hand, identification being confirmed on the basis of the shape pattern and number of scales on certain parts of the body. As in th other species, predominantly brown with yellow stripes or dark gre with yellow spots on their upperside forming a zigzag pattern. Howeve the present species only occurs on islands from which *P. biserialis* is absent, and although presen on the same islands as *A. slevini*, that species is considerably smaller.

E **3** Slevin's/Steindachner's Snake *Alsophis slevini*

Culebra slevini/steindachner

Length: up to 50 cm

Locally fairly common. Two named subspecies occur, mainly in the arid and shore zones: Slevin's Snake *slevini* (Fernandina, Isabela and Pinzón) and Steindachner's Snake *steindachneri* **3A** (Baltra, Rábida Santa Cruz and Santiago). Feeds by constricting its prey, although slightly venomous.

IDENTIFICATION: Virtually impossible to identify unless examined in the hand, identification being confirmed on the basis of the shape, pattern and number of scales on certain parts of the body. As in the other species, predominantly brown with yellow stripes or dark grey with yellow spots on the upperside forming a zigzag pattern. However, the present species only occurs on islands from which *P. biserialis* is absent and although present on the same islands as *A. dorsalis*, that species is considerably larger.

4 Yellow-bellied (or Pelagic) Sea Snake *Pelamis platurus* Culebra de Mar

Length: up to 85 cm

Migrant; regularly recorded at sea in Galápagos, particularly during El Niño years. Highly venomous.

IDENTIFICATION: Entirely marine. Black above and yellow below; flattened yellow tail with large black spots.

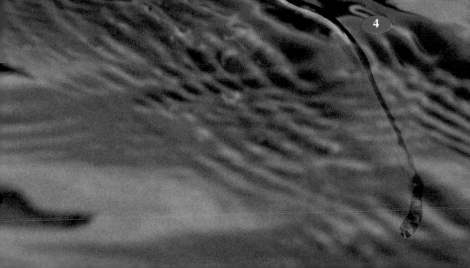

THE MAMMALS OF GALÁPAGOS

Introduction

In total, 32 indigenous species of mammal have been recorded in Galápagos in recent times. This excludes domesticated species which have become feral (dogs, cats, pigs, goats, donkeys, horses and cattle) and introduced rodents (rats and mice). Six species are endemic to the archipelago, four of which are confined to single islands. The great majority of the species recorded, 25 in total, are cetaceans (whales and dolphins). The mammals that occur in Galápagos can be divided into four 'types'. These are shown in the following table which indicates the number of species recorded and provides a summary of their status.

Type (Plate No.)	Species Recorded	Status		Endemic Species	Endemic Subspecies
		Residents	Migrants		
Sea lions (45)	2	2		1	1
Rodents (46–47)	4	4		4	
Bats (47)	2	2			1
Whales and dolphins (48–53)	25		25		

The types of mammal

This part of the book provides an introduction to each of the four 'types' of mammal that occur in Galápagos. It aims to aid initial identification, and the text for each of the types is cross-referenced to the relevant plate(s) that follow.

Galápagos Fur Seal

SEA LIONS   Family: Otaridae
2 species recorded (residents)   (Plate 45)

Endemic species: GALÁPAGOS FUR SEAL

Endemic subspecies: Galápagos Sea Lion

Sea lions and fur seals are the most conspicuous mammals in Galápagos. They are unmistakable, readily distinguished by their large size, characteristic shape and webbed flippers. Seal lions and fur seals feed at sea, spending the rest of the time on land in the shore zone.

Large Fernandina Rice Rat

RODENTS   Family: Muridae
4 indigenous species recorded   (Plates 46 and 47)
(plus introduced rats and mice)

Endemic species: LARGE FERNANDINA RICE RAT; SANTA FÉ RICE RAT; SANTIAGO RICE RAT; SMALL FERNANDINA RICE RAT

Before the arrival of man, the only species of rodent to occur in Galápagos were the rice rats. There were formerly seven species, all endemic to the archipelago, but the introduction of the ubiquitous Ship Rat is believed to have led to the extinction of three of these species. Rodents are small to medium-sized furry mammals with long scaly tails, pointed faces and short, rounded ears.

Galápagos Red Bat

BATS   Family: **Vespertilionidae**
2 species recorded (indigenous residents)   **(Plate 47)**
Endemic subspecies: Galápagos Red Bat
Bats are unmistakable, small, furry, nocturnal, flying mammals, their wings being formed of a membrane of skin between the elongated digits of their 'hands' and their ankles.

WHALES and DOLPHINS (CETACEANS)

Whales and dolphins (collectively known as cetaceans) are medium-sized to extremely large marine mammals. A total of 81 species is currently known to man and 25 of these have been reliably recorded in Galápagos.

Whilst our knowledge of the abundance, distribution and identification of cetaceans has improved tremendously since the early 1980s, there is still much to learn. Indeed, virtually nothing is known about some of the species recorded in Galápagos. This book summarises the information gathered to date on the key features for the identification of all of the cetacean species known to occur in the archipelago. Because cetacean identification is often a process of elimination, as much information as possible is given on all of the species. However, it should be borne in mind that some identification features are based on a limited number of sightings.

The information given on each species is based on our knowledge thus far, but do not be surprised if you experience something which does not fit the description in this book. Cetaceans are a constant joy to watch, not least because they are full of surprises. Just when we think it is safe to assume that Fin Whales never raise their tail flukes on diving and Pygmy Sperm Whales will not approach boats, they surprise us by doing just that! The important message here is that this section attempts to 'guide' the reader towards identification by illustrating several diagnostic features for each species – but it should not be considered definitive.

How to search for and identify cetaceans

The first challenge facing any whale watcher is finding a cetacean to identify! For this, all that is required is a pair of binoculars, a viewing platform (headland or boat) and some patience. Cetacean watching is easiest in relatively calm sea conditions, preferably with little swell and few or no white caps. Generally it is best to stand on the highest part of the platform and, if on a boat, try to look forward in front of the bow. Scan the water with the naked eye to cover the widest area and occasionally use binoculars for more distant searching. Keep searching the same area even if you think there is nothing there. Some cetaceans can dive for an hour or more.

Perhaps the most important point is to keep concentrating and remain patient. Most cetaceans are predators at the top of the food chain and, like terrestrial mammalian predators, they are not very abundant. This is all part of the excitement of whale watching, knowing that no matter how long the wait, something could pop up anywhere at any moment. Most surfacing cetaceans create their own white water so, if the sea is calm, a dark shape, a sudden splashing, or a blow is often the first sign of a sighting.

Having seen a shape in the water, the next task is to identify it. Identifying whales and dolphins at sea can be challenging for a number of reasons. The views are often brief, the animal may show little of its body above the water, and the state of the sea, weather and the sun's glare can make it difficult to establish colour, size and shape. These problems should not put off the amateur whale watcher. Whales and dolphins are magnificent and exciting creatures to watch and, like all field crafts, experience will allow the observer to identify species which seemed nearly impossible at first. However, both expert and novice must accept that because of the above difficulties, a relatively high proportion of sightings will be ascribed to one of the 'probable', 'possible' or 'haven't got a clue' categories.

The golden rule with cetacean identification is to try and pick out as many features as possible. Many species look quite similar and a number of characteristics are required for a confident identification. For example, Fin Whales can be identified by the asymmetrical coloration of the lower jaw. Unfortunately, most views at sea do not allow this feature to be noted. However, a combination of the tall straight blow, appearing before the dorsal fin as the animal surfaces, the long back, and the sloping fin combine to eliminate, with experience, the other large whales.

Having sighted a cetacean, the following features should be noted in approximate order of importance:

Size	Judging size and distance is very difficult at sea so try to compare with seabirds or boats. Remember that only part of an animal's back may be visible at any one time.
Blow	Is it visible? If so, size, shape and angle are important. Remember that the wind can affect these features.
Dorsal fin	Check its position along the back, and its size and shape.
Coloration and patterning	Colour is a variable and unreliable feature at sea and often the observer is reduced to describing coloration as 'light' or 'dark'. However, colours and, more importantly, patterns, are essential in identifying some species.
Behaviour	Behavioural traits can act as important supporting evidence for eliminating one species in favour of another. Such traits include fluking, porpoising, breaching, logging and spy-hopping.

Blow

Breaching – *lifting body out of water and splashing down on surface*

Spy-hopping – *head lifted and held out of the water*

Fluking – *tail flukes raised before a dive*

Cetacean topography

It is helpful to have some knowledge of the structure of a cetacean and the names of the main body parts. Cetaceans also exhibit some characteristic behaviours which have specific names. Familiarity with these names will help in understanding the text on each species and will make it easier to discuss identification with other whale watchers. The terms used in this book for the various parts of a cetacean are shown on the following illustrations of Blue Whale and Striped Dolphin.

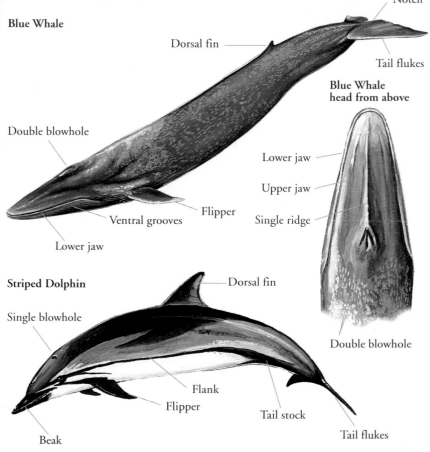

Blue Whale

Notch

Dorsal fin

Tail flukes

Double blowhole

Blue Whale head from above

Lower jaw

Upper jaw

Single ridge

Flipper

Ventral grooves

Lower jaw

Double blowhole

Striped Dolphin

Dorsal fin

Single blowhole

Flank

Flipper

Tail stock

Tail flukes

Beak

Logging – *lying motionless at the surface*

Porpoising – *raising part or all of body clear of the water while travelling*

The groups of cetacean

The cetaceans that have been recorded in Galápagos can be divided into five groups (which broadly equate to Families). This part of the book provides an introduction to each of these groups and aims to aid initial identification. It is cross-referenced to the relevant plate(s) and accompanying text which covers all of the species.

Blue Whale

RORQUAL WHALES   Family: **Baleanopteridae**
All 6 of the world's species recorded   (**Plates 48–49**)

The rorqual whales are large to very large, and include the biggest animals on earth. They differ from all other cetaceans in the region because they do not possess teeth. Instead, their upper jaws are lined with bony 'comb-like' plates called baleen which filter out small fish or zooplankton as the whale engulfs enormous quantities of seawater whilst swimming along. Rorquals have a double blowhole (single in toothed cetaceans), placed centrally on top of the head, a pleated throat capable of great expansion whilst feeding, a 'U'- to 'V'-shaped flattened head and a streamlined body for fast swimming.

Sperm Whale

SPERM WHALES Families: Kogidae & Physeteridae
3 species recorded   (**Plates 49–50**)

Although the sperm whales include two families and the largest and smallest species of toothed whale, their basic body plans are very similar. They have a characteristic large blunt head, low underslung jaw, and a blowhole positioned to the left of the centre of the head. Sperm whales are deep divers, capable of staying below for considerable periods of time as they hunt for squid and fish.

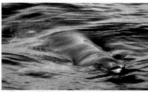
Blainville's Beaked Whale

BEAKED WHALES   Family: **Ziphidae**
3 species recorded   (**Plate 50**)

These medium-sized whales live in deep water and are rarely seen at sea, making them one of the least studied mammal groups on earth. As the name suggests, they all possess a distinctive protruding beak. Their bodies are generally slim and streamlined, often with visible scarring. They have no notch in the tail and a smallish dorsal fin, positioned two-thirds of the way along the back and variable in shape. All members of the family are similar in shape and variable in colour, making identification at sea extremely difficult. The position of protruding teeth which only erupt in adult males is often the only reliable feature. If possible, any species encountered should be photographed in order to learn more of the identity and distribution of these mysterious whales.

Short-finned Pilot Whale

BLACKFISH   Family: **Delphinidae**
5 species recorded   (**Plate 51**)

The blackfish include the largest members of the dolphin family: the killer and pilot whales. They are predominantly black with conspicuous dorsal fins. The jaws contain many well-developed conical teeth, but the beak is small or lacking. Like other dolphins they are highly social, fast and acrobatic, often breaching, spy-hopping and lobtailing. They are extremely effective and powerful pack-hunters, able to take fish, squid, and in some cases, marine mammals.

DOLPHINS
8 species recorded

Family: Delphinidae
(Plates 52–53)

Dolphins are smaller than most whale species and possess streamlined bodies, usually with a prominent, centrally placed dorsal fin, and a protruding beak. Their bodies have various patterns, colours and spotting which are often key to identification. Dolphins are generally social, capable of great speed, breathtaking acrobatics and bow-riding. They prey on a great variety of fish, squid and other marine life.

Long-snouted Spinner Dolphin

The mammal plates

The species illustrated Most of the species of mammal recorded in Galápagos are illustrated in the following plates. The only species to have been omitted are the Pacific Bottle-nosed Whale, which has been reliably sighted and photographed in the region, but remains unidentified for certain as no 'type' specimen has ever been found, and the Ginkgo-toothed Beaked Whale which has never been photographed at sea.

The species order For ease of reference, the plates are ordered so that all the potential confusion species are together. The species are numbered sequentially and the text follows the order in which the species appear on the plate, from top to bottom. The species are numbered sequentially and the following annotations are used on the plates:

1 Adult (species 1)   **1j** Juvenile

1m Male   **1f** Female

Distribution maps A distribution map accompanies the text for the sea lions, rodents and bats. As with the maps used in the previous sections, three 'groups' of islands have been moved to save space (see page 33 for further details). The distribution of the species is shown in orange on the maps in this section, as in the 'dummy' map opposite.

English name The English names used in this book for the mammals are those most commonly used in recent literature.

Spanish name The Spanish names shown are those which are most regularly used in Galápagos.

Endemic species The names of species which are endemic to Galápagos are preceded by an **E**.

Status The first part of the text for each species provides information on its status. This includes details of whether it is an indigenous resident, introduced, or migrant or vagrant, and a summary of their breeding season(s) and habitat preferences. The names of endemic subspecies are given in this section where appropriate.

Measurements The maximum recorded length of each species is given; in the case of cetaceans, the range of lengths of adults is shown.

Identification This section provides a summary of the key identification features, differences between males and females being described where necessary. In the case of cetaceans, details of the shape of the blow, the angle of breaching, whether the tail fluke is raised preceding a deep dive and the normal group size are given where this information aids identification.

Behaviour Particular behavioural characteristics which help in the identification of a species are highlighted where appropriate.

■ Breeding range (regular)

• Colony

■ Regularly seen at sea

▨ Occasionally seen at sea

□ Rarely seen at sea

131

e **1** Galápagos Sea Lion   *Zalophus californianus*   Lobo Marino de Galápag

Length: up to *c.* 230 cm

Weight: males up to 250 kg; females up to 120 kg

Common; endemic subspecies *wollebacki* of the California Sea Lio considered by some authorities to be a full species. Found in the shore zo and occasionally seen at sea away from land. Population estimated to numb *c.* 50,000.

IDENTIFICATION: Large and conspicuous and rather thinl furred. ADULT MALE: Considerably larger than female, with a steep sloping forehead and raised crown, and a very thick neck when full mature. Fur dark brown, often appearing black when wet. ADUL FEMALE: Considerably smaller than adult male and much mor graceful, with a dog-like muzzle. Fur creamy brown, appearing dar brown when wet.

VOICE: A variety of rather high-pitched barking and yelping sound

BEHAVIOUR: Forms often large colonies on sandy beaches. Male are highly territorial, particularly during the mating season which runs from May to Decembe coinciding with the Garúa season, vigorously defending their territories against intruder. Males take up to ten years to become fully mature but are sexually active at about six or seven females are fully mature at between six and eight years but can be sexually active at three year old. Pups are born during the Garúa season, although colonies on different islands give birt at different times, often up to three months apart. Immature males form discrete group known as 'bachelor colonies'. Often very inquisitive and playful in the water and individual regularly engage in body-surfing, often in a large wave. Feeds during the day.

E **2** Galápagos Fur Seal   *Arctocephalus galapagoensis*   Lobo de los Pelo

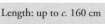

Length: up to *c.* 160 cm

Weight: males up to 75 kg; females up to 35 kg

Common. Found in the shore zone, mainly along rocky shores. Rarely seer at sea as feeds nocturnally. Population estimated to number *c.* 40,000.

IDENTIFICATION: Similar to Galápagos Sea Lion. Distinguishec by its considerably smaller size, much thicker fur, proportionall shorter and broader head, more pointed nose, large and slightl bulging eyes with a glazed expression, prominent ears which tend tc stick out from the head, and proportionally larger flippers. ADUL MALE: Considerably larger than female, with a thicker neck. Fu blackish to dark brown. ADULT FEMALE: Much smaller than adul male. Fur brown, sometimes with a slightly paler area on the nos and around the eyes.

VOICE: Less vocal than Galápagos Sea Lion, but the voice, although similar, is rather more guttural.

BEHAVIOUR: Forms small colonies on rocky shores, often lying in shady crevices during the heat of the day. Males are highly territorial, particularly during the mating season which runs from August to November, vigorously defending their territories against intruders. Male take five or six years to become fully mature; females are fully sexually mature at three years old. Most pups are born in October. Usually much shyer than the Galápagos Sea Lion. Feeds at night.

PLATE 4

E **1 Santiago Rice Rat** *Nesoryzomys swarthi* Rata Endemica de Santiag

Length: up to 35 cm

Only rediscovered on Santiago in 1997, having not been recorded sin 1906. Locally common in the arid zone to the north of the island. Althou nothing is yet known about its biology or ecology this is the only species rice rat which is known to have been able to compete successfully with t introduced Black Rat.

IDENTIFICATION: Similar in size and structure to the Larg Fernandina Rice Rat. The fur is dark brown, although the underpar are pale and the feet are white. The tail is about the same length the head and body.

BEHAVIOUR: Nocturnal.

E **2 Small Fernandina Rice Rat** *Nesoryzomys fernandinae*

Length: up to 22 cm

Raton Endemico de Fernandin

Only recently described, having been recorded only from owl pellets unt 1995 when the first live animals were trapped. Confined to Fernandin. Probably locally common, particularly at higher elevations, although nothin is yet known about its biology or ecology.

IDENTIFICATION: Considerably smaller than the Larg Fernandina Rice Rat, from which it is distinguished by its darke brownish fur and dark, rather than white, feet. The tail is slightl shorter than the length of the head and body.

BEHAVIOUR: Active at night.

E **3 Large Fernandina Rice Rat** *Nesoryzomys narboroughii*

Length: up to 35 cm

Rata Endemica de Fernandin

Locally common throughout the island, particularly in the arid zone Confined to Fernandina. Little is known about the biology or ecology o this species, although it is believed to breed principally during the warm wet season.

IDENTIFICATION: Considerably larger than the Smal Fernandina Rice Rat, from which it is distinguished by its paler greyish-black fur with pale underparts, and white feet. The tail is about the same length as the head and body.

BEHAVIOUR: Active at night and therefore rarely seen by visitors feeds on the ground.

E **4 Santa Fé Rice Rat** *Oryzomys bauri* Rata Endemica de Santa Fé

Length: up to 20 cm

Locally common, particularly in the arid zone. Confined to Santa Fé, where it is free from the depredations of Black Rat. Breeding takes place during the warm/wet season.

IDENTIFICATION: The Santa Fé Rice Rat is a small, brown rat with a pointed nose; long legs; long, pale but black-soled hindfeet; and pale underparts. They have large, rather bulging eyes and the ears are large and sparsely-haired. The tail is about the same length as the head and body, and is slender and naked.

BEHAVIOUR: Unafraid of humans but active principally at night and therefore rarely seen by visitors. When active, they spend their time on the ground.

PLATE 4

1 Hoary Bat   *Lasiurus cinereus*   Murciélago Escarchac

Size: Large (length of
forearm: *c.* 55 mm)

Locally common.

IDENTIFICATION: Short, blunt head and ears, no nose-leaf, ar
thickly-furred tail membrane. Considerably larger than Galápag
Bat. A large bat that tends to fly high off the ground, usually abo
c. 8 m, with a strong, fast flight and slow wingbeats. Fur light brow
with white 'frosting' and a white throat.

BEHAVIOUR: Like the Galápagos Bat, roosts in trees in shelter
places, especially in mangroves and along forest edges; usual
occurring singly. Occasionally forages around street lights in town

e **2** Galápagos Red Bat   *Lasiurus borealis*   Murciélago Vespertino de Galápag

Size: Small (length of
forearm: *c.* 40 mm)

Locally common resident; endemic subspecies *brachyotis*, which is consider
by some authorities to be a full species. Found in both the highlands ar
lowlands.

IDENTIFICATION: Short, blunt head and ears, no nose-leaf, an
thickly-furred tail membrane. Considerably smaller than Hoary Ba
Tends to fly relatively close to the ground, with rather fast win
beats. Fur bright rusty-orange on lower back; forequarters yellov
frosted with red; underparts yellowish.

BEHAVIOUR: A tree-roosting bat, usually occurring singly amon
foliage in sheltered places, especially along forest edges. Occasionall
forages around street lights in towns.

3 House Mouse   *Mus musculus*   Rato

Length: up to 9.5 cm

Introduced to many of the islands in Galápagos and included here to enable
comparison to be made with the rice rats. Nocturnal, mainly terrestrial.

IDENTIFICATION: Distinguished by its very small size and dark brownish-grey wit
slightly paler underparts. The tail is about the same length as the body.

4 Black (or Ship) Rat   *Rattus rattus*   Rata Negra

Length: *c.* 35 cm

Introduced to many of the islands in Galápagos,
where it has had a devastating effect on many
animal communities. Apart from on Santiago, no native rodents survive today
on islands where Black Rats are present. Although an introduced species, the
Black Rat is included in this book to enable a comparison to be made with the
rice rats. Nocturnal, mainly terrestrial, but very capable climbers.

IDENTIFICATION: Upperparts black or dark grey, or tawny-brown;
underparts grey. The tail is longer than the length of the head and
body and the whiskers are long, reaching as far as the shoulders.
The ears are of medium length.

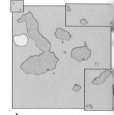

INTRODUCED SPECIES

   Black Rat and
   House Mouse

   Brown Rat

5 Brown (or Norway) Rat   *Rattus norvegicus*   Rata Doméstica

Length: *c.* 40 cm

Introduced to some of the islands where it is found mainly in areas of human
habitation. Although an introduced species, the Brown Rat is included in
this book to enable a comparison to be made with the rice rats. Nocturnal, mainly terrestrial.

IDENTIFICATION: Similar to tawny-brown form of Black Rat but generally larger, with
tail shorter than the length of the head and body, shorter whiskers that reach only as far as the
ears, and proportionally shorter ears.

PLATE 4

❶ Minke Whale *Balaenoptera acutorostrata*

Rorcual Aliblanc

Adult Length: 7–10 m
Blow: Small, vertical and bushy but usually not visible
Breaching: Variable angles
Deep dive: Tail flukes not raised
Group size: 1–2

Rare offshore.

IDENTIFICATION: The smallest of its family, simil in size and shape to the beaked whales but its point snout, with a single ridge on the upper surfac distinguishes it from toothed cetaceans. Differs fro other rorquals in its small size, lack of a tall blow, a large dorsal fin in relation to body length. Like the S Whale, the dorsal fin usually appears at the same tin as the blowhole on surfacing. Many individuals show white flipper band which is diagnostic.

❷ Sei Whale *Balaenoptera borealis*

Rorcual Bore

Adult Length: 12–16 m
Blow: Tall, thin and vertical – less robust than Fin Whale
Breaching: Seldom, generally rising at a low angle
Deep dive: Does not raise tail flukes
Group size: 1–2, sometimes more when feeding

Rare offshore.

IDENTIFICATION: Very similar to Bryde's Wha both in size and appearance, but the single ridge o top of the head distinguishes it at close range. Diffe from Minke Whale in size and presence of tall, visibl blow, and from Fin and Blue Whales by surfacin sequence and taller, more sickle-shaped fin. O surfacing, dorsal fin breaks the surface at the same tin as the blowhole.

❸ Bryde's Whale *Balaenoptera edeni*

Rorcual Tropica

Adult Length: 11–15 m
Blow: Tall, thin and vertical
Breaching: Usually leaves water at a steep angle
Deep dive: Unlike Sei Whale, often arches tail stock before dive; does not raise tail flukes
Group size: 1–2, sometimes more when feeding

Frequent inshore and offshore.

IDENTIFICATION: Easily confused with the Se Whale due to its similar size, shape, and sickle-shape dorsal fin. However, it possesses three lateral raise ridges on top of its head, a feature which distinguishe it from all other rorquals which only have a single ridg The surfacing sequence is also unique, with th blowhole disappearing from view just before the fin appears. This feature alone should not be used fo identification.

❹ Fin Whale *Balaenoptera physalus*

Rorcual Común

Adult Length: 18–26 m
Blow: Tall column, thicker and higher than Sei and Bryde's Whales but smaller than Blue Whale
Breaching: Variable angles, huge splash
Deep dive: Tail flukes not raised
Group size: 1–2, sometimes more when feeding

Rare offshore.

IDENTIFICATION: Second only to the Blue Whale in size, Fin Whales are similar to the other large rorquals. At close quarters, the asymmetrical pigmentation of the lower jaw is diagnostic. The left lower lip is dark, whilst the right is white. Surfacing sequence is distinctive: appearance of blowhole precedes a rolling back, followed by the small but distinct sloping dorsal fin.

PLATE 4

1 Blue Whale *Balaenoptera musculus*

Ballena Az[...]

Adult Length: 24–30 m

Blow: Largest of all: an enormous vertical column up to 10m tall

Breaching: Only young known to breach, usually at 45° angle

Deep dive: Sometimes raises tail flukes

Group size: 1–2, sometimes more when feeding

Rare offshore.

IDENTIFICATION: The Blue Whale is the large[...] animal on the planet, but its size is not the easiest featu[...] with which to distinguish it. Instead, concentrate o[...] surfacing sequence, fin size, and coloration. Confusio[...] is most likely with the Fin Whale but on surfacing [...] Blue Whale exhales an even larger blow. The head the[...] disappears to reveal a long rolling back before, finall[...] a tiny, stubby dorsal fin appears just before the anim[...] sinks below the surface. Body colour is bluish-grey bu[...] unlike Fin Whale, it is usually covered with pale mottlin[...]

2 Humpback Whale *Megaptera novaeangliae*

Jorobad[...]

Adult Length: 11–15 m

Blow: Variable; tall, vertical and bushy

Breaching: Yes, usually landing on back

Deep dive: Body generally arches high before broad tail flukes are raised to reveal variable pale underside to tail

Group size: 1–3, sometimes more when feeding

Occasional inshore and offshore.

IDENTIFICATION: Although similar to the othe[...] rorquals in size, the Humpback Whale is one of th[...] most distinctive species. Its scientific name means 'Big[...] winged New Englander', a description referring to it[...] huge flippers which are over 3m long. Surfacin[...] sequence is slow, the head is covered in raised knob[...] and the body is robust and bulky, with a low, slopin[...] and broad-based dorsal fin. The back behind the fin[...] like the Sperm Whale, shows a series of knuckles along[...] its ridge.

3 Sperm Whale *Physeter macrocephalus* (formerly *P. catodon*)

Cachalote[...]

Adult Length: 11–18 m

Blow: Single blowhole on left side of head; blow angled forward and to the left

Breaching: Frequent, most of body leaves water

Deep dive: Often arches body; usually raises dark, broad, triangular tail flukes vertically

Group size: Group size: 1–20, sometimes more

Frequent offshore.

The abundance of Sperm Whales in this region made it the focus of the whaling industry in the nineteenth century, and of extensive research in recent years. The seas around Galápagos form important feeding and breeding grounds for this species. Since the late 1980s numbers have declined from a few hundred to generally 10–20 in the 1990s. The reason for this change is unknown.

IDENTIFICATION: The Sperm Whale is the largest toothed whale, overlapping with several rorquals in size. However, its squarish head, blow, and wrinkled skin make it very distinctive. At the surface it generally shows a low-lying, straight back. There is a small hump instead of a fin, behind which are a series of knuckles running as far as the tail.

① Dwarf Sperm Whale · *Kogia simus* · Cachalote Ena...

Adult Length: 2·1–2·8 m
Blow: Blowhole displaced slightly to the left; blow low and faint
Breaching: Leaps vertically
Deep dive: Tail flukes not raised
Group size: 1–10

Occasional offshore and inshore.

IDENTIFICATION: Very similar to Pygmy Sper... Whale (see identification notes for that species) b... marginally smaller and best identified by its slightly larg... dorsal fin positioned centrally on the back. Identificati... remains very difficult without exceptionally good vie... and consequently observations at sea are often record... as Pygmy/Dwarf Sperm Whale.

② Pygmy Sperm Whale · *Kogia breviceps* · Cachalote Pigm...

Adult Length: 2·7–3·7 m
Blow: Blowhole displaced slightly to the left; blow low and faint
Breaching: Leaps vertically
Deep dive: Tail flukes not raised
Group size: 1–6

Rare offshore.

IDENTIFICATION: The small size of the Pygmy Sper... Whale means that it is most likely to be mistaken for... dolphin, although its blunt, squarish head, robust bo... small dorsal fin and behaviour are distinctive. However... is extremely similar to the Dwarf Sperm Whale. Both spec... are typically seen travelling slowly or hanging motionless... the surface and on diving, simply sink below the surfa...

rather than rolling forward. Both are also usually shy of boats and, when disturbed, excrete a reddis... brown ink in defence. The Pygmy Sperm Whale is slightly larger but the size and position of t... dorsal fin is the key feature, being smaller and positioned further than halfway along the back.

③ Blainville's Beaked Whale · *Mesoplodon densirostris* · Zifio de Blainvil...

Adult Length: 4–6 m
Coloration: Blue-grey to orange-brown
Blow: Small blow projects forward, sometimes visible
Breaching: Unknown
Group size: 1–6

Occasional offshore.

IDENTIFICATION: Very similar in shape and size ... other beaked whales of the genus *Mesoplodon*. Identificatio... requires close views of the beak which appears first as th... animal surfaces. The forehead is low and flat leading to... shortish beak, behind which both sides of the lower jaw a... distinctly arched towards the base. Unlike Gingko-toothe... Beaked Whale the arch is rounded. Mature males display... large single tooth, irrupting from the centre of each arch...

Ginkgo-toothed Beaked Whale · *Mesoplodon ginkgodens* · Zifio Japon...

Adult Length: 5 m
Coloration: Uniformly dark
Blow: Unknown
Breaching: Unknown
Group size: Unknown

Status unknown: 1 stranding, on Genovesa, in June 1970.

IDENTIFICATION: Similar in body shape to Blainville... Beaked Whale, so close views of the head are required fo... identification. The forehead slopes smoothly into the uppe... jaw, forming a prominent beak. Both sides of the lower jaw... rise steeply halfway along the beak before levelling out a... eye height. This gives the beak an angled, arched appearanc...

which, unlike Blainville's, does not protrude above the upper jaw. In older males, a single tooth (th... base of which is shaped like the leaf of a Ginkgo Tree) irrupts from the tip of each arch.

④ Cuvier's Beaked Whale · *Ziphius cavirostris* · Zifio Comú...

Adult Length: 5–7 m
Coloration: Grey-brown to brick-red
Blow: Bushy, slightly angled forward, sometimes visible
Breaching: Leaps almost vertically
Group size: 1–8

Occasional offshore.

IDENTIFICATION: Larger and more robust than the... beaked whales of the genus *Mesoplodon*. The body i... somewhat sausage-shaped as it rolls slowly at the surface... revealing a small, triangular or falcate dorsal fin situated... two-thirds of the way along the back. The head shape... is distinctive, the forehead sloping gently down to a... short beak, described by some as being like that of a... goose. In mature males, two small teeth protrude from... the tip of the lower jaw. The head and upper back of... adults often becomes pale cream in colour.

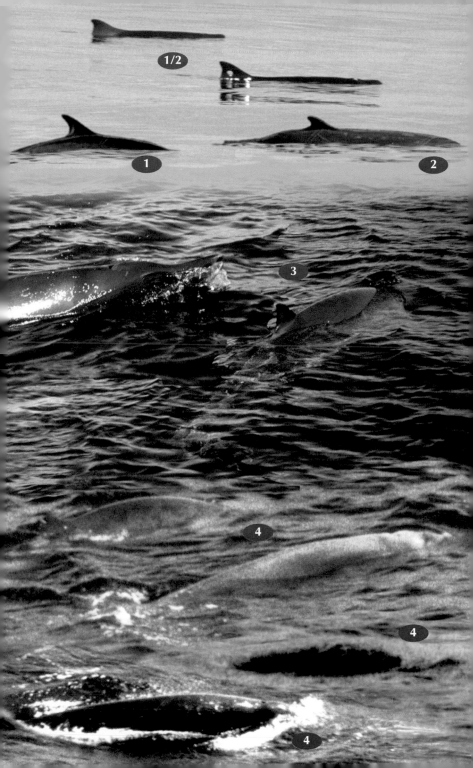

① Killer Whale (or Orca)   *Orcinus orca*

Adult Length: 5–9 m
Blow: Tall and bushy
Breaching: Leaps vertically
Group size: Family groups of 2–30

Frequent inshore and offshore.

IDENTIFICATION: The largest of the dolphins a
perhaps the most distinctive cetacean of all. The striki
black body, white eye-patch and prominent dorsal
are unmistakable. Dorsal fin size and shape varies w
age and sex. Adult males possess triangular fins whi
can reach up to 2m in height. Females and immatures have smaller, sickle-shaped fins. Kno
to hunt other marine mammals including Sperm Whales and Bryde's Whales in Galápagos.

② False Killer Whale   *Pseudorca crassidens*   Falsa O

Adult Length: 4–6 m
Blow: Inconspicuous and bushy
Breaching: Frequent, various angles
Group size: 2–200

Occasional inshore and offshore.

IDENTIFICATION: Similar in shape to Melon-head
Whale and Pygmy Killer Whale but much larger and mc
powerful. Confusion is perhaps most likely with fema
Orca or Short-finned Pilot Whale. With good views t
all-dark body distinguishes it from the former, and
slender, rounded head and body, with a dolphin-like dorsal fin, separates it from the latter. T
position of the dorsal fin also differs from that of the Pilot Whale, being set further back at th
mid-point of the back. At the surface False Killer Whales are often highly active swimmers, lifti
their whole heads and bodies out of the water whilst travelling. Known to attack other cetacean
including Sperm Whales, in Galápagos.

③ Short-finned Pilot Whale   *Globicephala macrorhyncus*   Calderon Tropic

Adult Length: 3·5–6·5 m
Blow: Distinctly bushy
Breaching: Occasional, various angles
Group size: Family groups of
2–50

Occasional inshore and offshore.

IDENTIFICATION: The Short-finned Pilot Whale is or
of the easiest species to identify due to its characteristic shap
and behaviour. On surfacing, the bulbous, rounded hea
precedes a robust body with a dorsal fin set well forwar
Behind the dorsal fin is a pale 'saddle patch' followed by
long back and tail stock. In both sexes the dorsal fin is ver
broad at the base, but this is especially so in adult males which have a large, flag-shaped fin. Becaus
they tend to feed at night, most of the day is spent either travelling at a leisurely pace or logging.

④ Melon-headed Whale   *Peponocephala electra*   Calderon Pequeñ

Adult Length: 2–2·7 m
Blow: Generally not visible
Breaching: Occasional
Group size: Generally 50+

Rare offshore.

IDENTIFICATION: This small, slender species is aki
in size and shape to most dolphins. However its all
black coloration and lack of a prominent bea
distinguish it. Confusion is most likely with the Pygm
Killer Whale which is extremely similar. Close views i
good light are required but there are several subtle differences. Perhaps most diagnostic i
head shape which is slightly pointed or 'melon-shaped'. Other differences include slimme
and more sharply pointed flippers, lack of a dark cape, and a short but distinct beak, ofte
present on females and juveniles.

⑤ Pygmy Killer Whale   *Feresa attenuata*   Orca Pigmea

Adult Length: 2–2·7 m
Blow: Generally not visible
Breaching: Occasional
Group size: Generally 1–50

Rare offshore.

IDENTIFICATION: Only likely to be confused with
the Melon-headed Whale, as its all-dark, grey-black
coloration and lack of a beak distinguishes it from the
dolphins. The similarity with Melon-headed Whale
means that close views are required for positive
identification. Head shape is an important feature to note, being distinctly rounded in Pygmy
Killer Whales. Another significant difference is the black cape, which reaches its maximum
width below the dorsal fin, and is distinctive in some animals. Some individuals may also
have white chins. The flippers are rounded at the tips.

PLATE 5

1 Common Dolphin   *Delphinus delphis*

Adult Length: 1·7–2·4 m
Behaviour: Highly active, fast swimmer capable of impressive acrobatics
Group size: 1–2,000

Frequent offshore, occasional inshore.

IDENTIFICATION: This small and acrobatic dolphi is very similar in shape to the Striped Dolphin an Pantropical Spotted Dolphin, so identification in poc light can be difficult. In good light this species is easil identified by its coloration and patterning. The dors. fin and cape are black, with the cape dropping to form 'V'-shape on the flanks below the dorsal fin. The flank is coloured in a criss-crossed patter with yellow-buff from the eye to the dorsal fin, and with grey between the dorsal fin and the tai The base of the dorsal fin is often pale and a black line runs between the beak and the flippe

2 Striped Dolphin   *Stenella coeruleoalba*

Adult Length: 1·8–2·5 m
Behaviour: Perhaps even more acrobatic than the Common Dolphin, sometimes breaching as high as 7m. Often cautious around boats, swimming in a tightly-packed pod
Group size: Generally 10–500

Occasional offshore.

IDENTIFICATION: Due to similarities in size, shap and behaviour with Common and Pantropical Spotte Dolphins, pattern and colour are key to the identificatio of this small, slender cetacean. A thin dark body-strip runs along a pale flank from behind the eye to the tai stock, and another stripe runs from the eye to the flippe giving the Striped Dolphin its name. Surprisingly, this i not the most reliable identification feature because both stripes remain below the surface when individuals are not porpoising. Fortunately, the dark grey cape on the upper back is distinctive, being dissected by a pale strip angled backwards toward: the dorsal fin.

3 Long-snouted Spinner Dolphin   *Stenella longirostris*   Estenela Giradora

Adult Length: 1·5–2·2 m
Behaviour: One of the most acrobatic of all dolphins, capable of twisting its body longitudinally to spin up to 7 times in a single leap
Group size: Generally 5–200

Occasional offshore.

IDENTIFICATION: This small, streamlined dolphin with a long beak and lack of patterning is only likely to be confused with the Pantropical Spotted Dolphin. There are three distinct races in the eastern tropical Pacific which are likely to occur in Galápagos and these vary in colour and beak length. Colour varies from uniform dark grey to bicoloured or tricoloured shades of grey, fading paler towards the belly. In mature males the dorsal fin is often an erect triangular shape or angled slightly forward, giving the impression that the fin is on backwards! Older males also develop a distinct anterior bulge under the tail stock. This is unique and clearly visible in leaping dolphins. Spinner Dolphins are often characterised by their crazy spinning behaviour.

4 Pantropical Spotted Dolphin   *Stenella attenuata*   Delfín Moteado

Adult Length: 1·7–2·4 m
Behaviour: Highly energetic and acrobatic
Group size: Generally 50–1,000

Occasional offshore.

IDENTIFICATION: Very similar to Common and Striped Dolphins in shape but lacks flank markings. Instead, the dark dorsal fin and cape give way to lighter grey flanks and a pale grey belly (the belly is white in Common and Striped Dolphins). A dark band runs between the beak and the dark flipper. Calves are unspotted at birth but become more spotted with age until, in adulthood, the body is completely covered. Often occurs in mixed pods with Long-snouted Spinner Dolphins but are easily told apart by the spotting and curved dorsal fin.

① Risso's Dolphin *Grampus griseus*

Delfín G

Occasional offshore.

Adult Length: 2·6–3·8 m

Behaviour: Can be energetic but more frequently engages in slow travel or logging. Displays many behaviours including spy-hopping, lobtailing and breaching

Group size: 1–500

IDENTIFICATION: The distinctive colour and sha of the Risso's Dolphin means that it is unlikely to mistaken given good views. In shape it is perhaps mc reminiscent of a blackfish than a dolphin due to i large size, bulbous head, short indistinct beak ar prominent falcate dorsal fin. However, the extensi scarring and whitish skin are diagnostic (although similar pattern is exhibited by mature Cuvier's Beak

Whales). Born pale grey, the body darkens until maturity, then lightens to a creamy white. I this time the body tends to be covered in scratch marks from infighting and scars inflicted I their squid prey. Adults become continually paler with age but the dorsal fin and adjacer back often remain dark.

② Fraser's Dolphin *Lagenodelphis hosei*

Delfín de Fras

Rare offshore.

Adult Length: 2–2·6 m

Behaviour: Powerful swimming style, often leaving the water in a burst of spray

Group size: 100–500

IDENTIFICATION: Although amongst the lea: known of all dolphins, Fraser's Dolphin has a distinctiv shape. It appears robust towards the head, slende behind the small, curved dorsal fin, and has a sho beak. These characteristics are intermediate between th

genera *Lagenorhyncus* and *Delphinus*, hence the species' scientific name. Flank patterning i also very distinctive. Some individuals, especially males, have a thick, dark lateral body-strip which runs from the eye backwards and is bordered by a pale stripe above and a pale bell below. Another dark line runs from the beak to the flippers. The strength of the stripe varie greatly from individual to individual, but at least some dolphins in a pod generally sho strong markings.

③ Rough-toothed Dolphin *Steno bredanensis*

Stenc

Rare: 1 stranding, at Tortuga Bay, Santa Cruz Island in 1964

Adult Length: 2·1–2·6 m

Behaviour: Sometimes shows a characteristic swimming pattern described as 'skimming across the surface'

Group size: 1–50

IDENTIFICATION: Wholly dark grey body make: confusion with Spotted, Spinner and Bottlenose Dolphins and the smaller blackfish possible. The narrow elongated beak is the most distinctive feature, tapering gradually from the forehead to give a cone-shaped impression. All other dolphins have a clear demarcation

between the forehead and beak. The beak also often shows whitish or pinkish patches, particularly around the lips, but this may extend to a white throat. Below the tall prominent dorsal fin is a narrow dark cape, a feature which is lacking in the uniformly coloured Bottlenose Dolphin.

④ Bottle-nosed Dolphin *Tursiops truncatus*

Delfín Mular

Frequent inshore and offshore.

Adult Length: 1·9–3·90 m

Behaviour: Highly active, capable of great speed and amazing acrobatics

Group size: 1–25

IDENTIFICATION: Due to its popularity in aquaria and with the media, this is perhaps the most familiar of all the dolphins. It is also one of the most distinctive. Bottle-nosed Dolphins are large, robust animals which move powerfully through the water. Their size and plain

grey coloration, paler below but with no distinct flank markings, set them apart from all but the Rough-toothed Dolphin. However, the distinct bulging forehead and short, stubby beak distinguish it from Rough-toothed.

CHECKLIST OF THE REGULARLY OCCURRING SPECIES

WITH SUMMARY OF HABITAT PREFERENCES AND DISTRIBUTION

KEY TO THE TABLE

The symbols used for species' status:

E   species endemic to the Galápagos
e   subspecies endemic to the Galápagos
r   resident species
m   regular migrant
i   introduced species
?   status uncertain

The symbols used in the habitat preference columns:

■   favoured habitat
▪   sometimes recorded

The symbols used in the islands columns:

●   breeds
•   occurs but does not breed or is uncommon
＊   breeds only on satellite islands
?   status uncertain
▨   occurs in the waters around the island
▬   migrant, may be found in coastal waters
▬   migrant, may be found in suitable habitat on any island

ssp.   denotes endemic subspecies where more than one occurs in the Galápagos

		Habitat preference						Islands										
		Open sea	Rocky islets	Shore zone	Arid zone	Transition zone	Humid zone	Española (Hood)	Floreana (Charles)	San Cristóbal (Chatham)	Santa Fé (Barrington)	Santa Cruz (Indefatigable)	Baltra & Seymour	Pinzón (Duncan)	Isabela (Albemarle)	Fernandina (Narborough)	Santiago (James)	Genovesa (Tower)
	BIRDS																	
✓	**SEABIRDS**																	
E	Albatross, Waved	■			■			●										
e	Booby, Blue-footed	■	■	■	■			●	●	●	●	＊	＊		●		●	•
r	Booby, Nazca	■	■	■	■			●	＊	●	＊				●		●	●
r	Booby, Red-footed	■	■	■	■					＊					●			●
E	Cormorant, Flightless (or Galápagos)	▪		■											●	●	●	
r	Frigatebird, Great	■	■	■	■			●	●	●			•		•		•	●
e	Frigatebird, Magnificent	■	■	■	■			＊	●	●		＊	●				●	
m	Gull, Franklin's	■	▪					▬ migrant ▬										
m	Gull, Laughing	■	▪					▬ migrant ▬										
E	Gull, Lava	■	■	■				•	•	•	•	•	•	•	•	•	•	•
E	Gull, Swallow-tailed	■	■	■		▪		•	•	•	•	•	•	•	•	•	•	•
m	Jaeger, Pomarine	■			■			▬ migrant ▬										
e	Noddy, Common	■	■	■		▪		●	●	●	●	●	●	●	●	●	●	●
e	Pelican, Brown	■	■	■				●	●	●	●	●	●	●	●	●	●	●
E	Penguin, Galápagos	■	▪	■				•	•	•		•		•	●	●	●	•
e	Petrel, Dark-rumped (or Galápagos)	■					■		●	●		●			●		●	
e	Shearwater, Audubon's	■	■	■	▪			●	●	●	●	●	●	●	●	●	●	●
m	Shearwater, Sooty	■						▬ migrant ▬										
e	Storm-petrel, Elliot's (or White-vented)	■						▬ migrant ▬										
r	Storm-petrel, Madeiran	■	■	▪	■			●	＊	＊		＊				＊		●
m	Storm-petrel, Markham's	■						▬ migrant ▬										
e	Storm-petrel, Wedge-rumped	■	■	■						●					●			●
m	Tern, Common	■	▪					▬ migrant ▬										
m	Tern, Royal	■	▪					▬ migrant ▬										
r	Tern, Sooty	■	■	■				Darwin only										
r	Tropicbird, Red-billed	■	■	■		▪		●	●	●	●	●	●	●	●	●	●	●

150

Status	Name	Open sea	Rocky islets	Shore zone	Arid zone	Transition zone	Humid zone	Española (Hood)	Floreana (Charles)	San Cristóbal (Chatham)	Santa Fé (Barrington)	Santa Cruz (Indefatigable)	Baltra & Seymour	Pinzón (Duncan)	Isabela (Albemarle)	Fernandina (Narborough)	Santiago (James)	Genovesa (Tower)
WATERBIRDS																		
r	Crake, Paint-billed				·	■	·		●	●		●			●			
r	Egret, Cattle			■	·	■	·		●	●		●	·		●	·	·	
r	Egret, Great			■	·	■	·		·	●		●	·	·	●	●	●	
m	Egret, Snowy			■														
r	Flamingo, Greater			■					●			●			●		●	
r	Gallinule, Common (or Moorhen)			■						●		●			●	·		
e	Heron, Great Blue			■					●	●	·	●	·	·	●	·	●	
E	Heron, Lava (or Galápagos)			■				●	●	●	●	●	●	●	●	●	●	●
r	Heron, Striated			■	·					●		●			●		●	·
e	Night-heron, Yellow-crowned		·	■				●	●	●	●	●	●	●	●	●	●	●
e	Pintail, White-cheeked (or Galápagos)			■		■		·	●	●	·	●	·	·	●	·	●	·
E	Rail, Galápagos					·	■		·	·		●			●		●	
m	Teal, Blue-winged			■		■	■											
SHOREBIRDS																		
m	Dowitcher, Short-billed			■														
e	Oystercatcher, American		·	■				·	●	●	●	●	·	·	●	·	●	·
m	Phalarope, Red (or Grey)	■		·														
m	Phalarope, Red-necked	■		·														
m	Phalarope, Wilson's			■														
m	Plover, Black-bellied (or Grey)			■														
m	Plover, Semipalmated			■														
m	Sanderling			■														
m	Sandpiper, Least			■														
m	Sandpiper, Solitary					■	·											
m	Sandpiper, Spotted			■														
r	Stilt, Black-necked			■		■	·		●	●		·	·		●	·	●	
m	Surfbird			■														
m	Tattler, Wandering			■														
m	Turnstone, Ruddy			■														
m	Whimbrel			■		■	·											
m	Willet			■														
m	Yellowlegs, Lesser			■														
RAPTORS																		
E	Hawk, Galápagos			■	■	■	■		●			●		·	●	●	●	
m	Osprey	·		■	·													
m	Peregrine	·		■	·													
NIGHTBIRDS																		
e	Owl, Barn				■	■	■			●		●	·		●	●	●	
e	Owl, Short-eared			·	■	■	■	●	●	●	●	●	●	●	●	·	●	●
LARGER LANDBIRDS																		
r?	Ani, Groove-billed				·	■	·					?						
r	Ani, Smooth-billed				·	■	·		●			●			●		·	
r	Cuckoo, Dark-billed			·	■	■	■		●	●		●	·	·	●	·	●	
E	Dove, Galápagos		·	■	■	·		●	·	·	●	●	·	●	●	·	●	●
m	Kingfisher, Belted			■														
AERIAL FEEDERS																		
E	Martin, Galápagos			■	■	■	■	·	●	●	·	●	·	·	●	●	●	●
m	Swallow, Barn				■	■	■											

151

HABITAT PREFERENCE — Open sea, Rocky islets, Shore zone, Arid zone, Transition zone, Humid zone

ISLANDS — Española (Hood), Floreana (Charles), San Cristóbal (Chatham), Santa Fé (Barrington), Santa Cruz (Indefatigable), Baltra & Seymour, Pinzón (Duncan), Isabela (Albemarle), Fernandina (Narborough), Santiago (James), Genovesa (Tower)

✓	St	SMALLER LANDBIRDS	Open sea	Rocky islets	Shore zone	Arid zone	Transition zone	Humid zone	Española	Floreana	San Cristóbal	Santa Fé	Santa Cruz	Baltra & Seymour	Pinzón	Isabela	Fernandina	Santiago	Genovesa
	E	Finch, Cactus ssp. *abingdoni*			■	■	▪		Pinta only										
		ssp. *intermedia*			■	■	▪		●	●	●	●	●	●	?	●			
		ssp. *rothschildi*			■	■	▪	▪	Marchena only										
		ssp. *scandens*			■	■	▪	▪										●	
	E	Finch, Large Cactus ssp. *conirostris*			■	■			●										
		ssp. *darwinii*			■	■			Darwin and Wolf only										
		ssp. *propinqua*			■	■													●
	E	Finch, Large Ground		▪	■	■	▪	▪					●	●	●	●		●	●
	E	Finch, Large Tree ssp. *affinis*				▪	■	■								●	●		
		ssp. *habeli*				■	■	■	Pinta and Marchena only										
		ssp. *psittacula*				▪	■	■		●			●					●	●
	E	Finch, Mangrove			■											●	●		
	E	Finch, Medium Ground		■	■	■	■	▪		●	●	●	●	●	●	●	●	●	
	E	Finch, Medium Tree				▪	■	■		●									
	E	Finch, Sharp-beaked Ground ssp. *debilirostris*			▪	▪	■	■								●		●	●
		ssp. *difficilis*			■	■			also Pinta										●
		ssp. *septentrionalis*			■	■			Darwin and Wolf only										
	E	Finch, Small Ground		■	■	■	■	▪	●	●	●	●	●	●	●	●	●	●	●
	E	Finch, Small Tree ssp. *parvulus*			▪	▪	■	■		●	●		●	●	●	●	●	●	
		ssp. *salvini*			▪	▪	■	■			●								
	E	Finch, Vegetarian				▪	■	■		●	●		●		●	●	●	●	
	E	Finch, Warbler ssp. *becki*			■	■			Darwin and Wolf only										
		ssp. *bifasciata*			■	■							●						
		ssp. *cinerascens*			■	■			●										
		ssp. *fusca*			▪	■	■	■	Pinta and Marchena only										
		ssp. *luteola*			▪	■	■	■		●									
		ssp. *mentalis*			■	■	■	■											●
		ssp. *olivacea*			■	■	■	■					●	●	●	●	●	●	
		ssp. *ridgwayii*			▪	■	■	■		●									
	E	Finch, Woodpecker ssp. *pallidus*			▪	■	■	■							●	•		●	
		ssp. *productus*			▪	■	■	■								●	●		
		ssp. *striatipectus*			▪	■	■	■			●								
	E	Flycatcher, Large-billed (or Galápagos)			■	■	■	■	●	●	●	●	●	●	●	●		●	●
	e	Flycatcher, Vermilion ssp. *dubius*		▪	▪	■	■	■		●			●						
		ssp. *nanus*			▪	▪	■	■		●			●		●	●	●	●	
	E	Mockingbird, Charles (or Floreana)	■							※									
	E	Mockingbird, Chatham			■	■	■	▪			●								
	E	Mockingbird, Galápagos ssp. *bauri*			■	■													●
		ssp. *barringtoni*			■	■						●							
		ssp. *hulli*			■	■			Darwin only										
		ssp. *parvulus*		■	■	■	▪	▪							●	●		●	●
		ssp. *personatus*		▪	■	■	■	▪	also Pinta, Marchena and Rábida									●	
		ssp. *wenmani*			■	■			Wolf only										
	E	Mockingbird, Hood	■		■	■			●										
	e	Warbler, Yellow			■	■	■	■	●	●	●	●	●	●	●	●	●	●	●

	HABITAT PREFERENCE						ISLANDS										
		Open sea	Rocky islets	Shore zone	Arid zone	Transition zone	Humid zone	Española (Hood)	Floreana (Charles)	San Cristóbal (Chatham)	Santa Fé (Barrington)	Santa Cruz (Indefatigable)	Baltra & Seymour	Pinzón (Duncan)	Isabela (Albemarle)	Fernandina (Narborough)	Santiago (James)	Genovesa (Tower)

REPTILES

TORTOISE

		Open sea	Rocky islets	Shore zone	Arid zone	Transition zone	Humid zone	Esp	Flo	SCr	SFé	SCz	B&S	Pin	Isa	Fer	San	Gen
E	Tortoise, Galápagos ssp. *abingdoni*							Pinta only - **EXTINCT in the wild**										
	ssp. *becki*				▪	■	■								●			
	ssp. *chathamensis*				▪	■	■			●								
	ssp. *darwini*				▪	■	■										●	
	ssp. *elephantopus*				▪	■	■								●			
	ssp. *ephippium*				▪	■	■							●				
	ssp. *guntheri*				▪	■	■								●			
	ssp. *hoodensis*					■	■	●										
	ssp. *microphyes*				▪	■	■								●			
	ssp. *porteri*				▪	■	■					●						
	ssp. *vandenburghi*				▪	■	■								●			

TURTLES

		Open sea	Rocky islets	Shore zone	Arid zone	Transition zone	Humid zone	Esp	Flo	SCr	SFé	SCz	B&S	Pin	Isa	Fer	San	Gen
r	Turtle, Black (or Pacific Green)	■		■				●	●	●	●	●	●	●	●	●	●	●
m	Turtle, Hawksbill	■																
m	Turtle, Loggerhead	■																
m	Turtle, Olive Ridley	■																

GECKOS

		Open sea	Rocky islets	Shore zone	Arid zone	Transition zone	Humid zone	Esp	Flo	SCr	SFé	SCz	B&S	Pin	Isa	Fer	San	Gen
E	Leaf-toed Gecko, Baur's			■	■	■		●	●									
E	Leaf-toed Gecko, Galápagos			■	■	■						●	?	●	●	●	●	●
E	Leaf-toed Gecko, Rábida				■	■		Rábida only - possibly **EXTINCT**										
E	Leaf-toed Gecko, San Cristóbal				■	■				●								
E	Leaf-toed Gecko, Santa Fé				■	■					●							
r	Leaf-toed Gecko, Tuberculated				■	■				●								
E	Leaf-toed Gecko, Wenman				■	■		Wolf only										
i	*Gonatodes caudiscutatus*			Buildings						●								
i	*Lepidodactylus lugubris*			Buildings						●								
i	*Phyllodactylus reissi*			Buildings						●								

LIZARDS

		Open sea	Rocky islets	Shore zone	Arid zone	Transition zone	Humid zone	Esp	Flo	SCr	SFé	SCz	B&S	Pin	Isa	Fer	San	Gen
E	Lava Lizard, Española			■	■	■		●										
E	Lava Lizard, Floreana			■	■	■			●									
E	Lava Lizard, Galápagos			■	■	■		& Rábida		●	●	●			●	●	●	
E	Lava Lizard, Marchena			■	■			Marchena only										
E	Lava Lizard, Pinta			■	■			Pinta only										
E	Lava Lizard, Pinzón			■	■									●				
E	Lava Lizard, San Cristóbal			■	■					●								

IGUANAS

		Open sea	Rocky islets	Shore zone	Arid zone	Transition zone	Humid zone	Esp	Flo	SCr	SFé	SCz	B&S	Pin	Isa	Fer	San	Gen
E	Iguana, Land				■	■						●	●		●		●	
E	Iguana, Marine ssp. *albemarlensis*	▪	■	■	▪										●			
	ssp. *cristatus*	▪	■	■	▪							●						
	ssp. *hassi*	▪	■	■	▪				?		?	●	?	?				
	ssp. *mertensi*	▪	■	■	▪					●							●	
	ssp. *nanus*	▪	■	■	▪													●
	ssp. *sielmanni*	▪	■	■	▪			Pinta only										
	ssp. *venustissimus*	▪	■	■	▪			●										
E	Iguana, Santa Fé Land			■	■						●							

	HABITAT PREFERENCE						ISLANDS									
	Open sea	Rocky islets	Shore zone	Arid zone	Transition zone	Humid zone	Española (Hood)	Floreana (Charles)	San Cristóbal (Chatham)	Santa Fé (Barrington)	Santa Cruz (Indefatigable)	Baltra & Seymour	Pinzón (Duncan)	Isabela (Albemarle)	Fernandina (Narborough)	Santiago (James)
✓ **SNAKES**																
E Snake, Floreana			■	■	■			●								
—, Española			■	■	■		●									
—, San Cristóbal			■	■	■				●							
E Snake, Galápagos			■	■	■		& Rábida			●	●	●				●
—, Fernandina			■	■	■										●	
—, Isabela			■	■	■									●		
E Snake, Slevin's			■	■	■								●	●	●	
—, Steindachner's			■	■	■		& Rábida			●	●					●
m Snake, Yellow-bellied Sea	■															
MAMMALS																
SEA LIONS																
E Fur Seal, Galápagos		■	■	■			●	●	●	●	●	●	●	●	●	●
e Sea Lion, Galápagos	■	■	■	■			●	●	●	●	●	●	●	●	●	●
RICE RATS																
E Rice Rat, Large Fernandina				▪	■	■									●	
E Rice Rat, Santa Fé				▪	■					●						
E Rice Rat, Santiago				▪	■											●
E Rice Rat, Small Fernandina				▪	■	■									●	
INTRODUCED RODENTS																
i Mouse, House			■	■	■	■		●	●		●	●	●	●		●
i Rat, Black (or Ship)			■	■	■	■	●	●	●		●	●	●	●		●
i Rat, Brown (or Norway)	Towns & habitations								●		●			●		
BATS																
r Bat, Hoary			■	■	■	■		●	●		●			●		●
e Bat, Galápagos Red			■	■	■	■			●		●					
WHALES																
? Whale, Blainville's Beaked	■															
r Whale, Bryde's	■															
? Whale, Cuvier's Beaked	■															
m Whale, False Killer	■															
m Whale, Humpback	■															
r Whale, Killer (or Orca)	■															
r Whale, Short-finned Pilot	■															
r Whale, Sperm	■															
DOLPHINS																
r Dolphin, Bottle-nosed	■															
m Dolphin, Common	■															
m Dolphin, Long-snouted Spinner	■															
m Dolphin, Pantropical Spotted	■															
m Dolphin, Risso's	■															
m Dolphin, Striped	■															

GLOSSARY

Blackfish	General term for the larger members of the dolphin family.
Blowhole	The nostril of a cetacean, situated on the top of its head. Toothed whales have a single nostril whilst baleen whales have a double.
Bow-riding	Riding on the pressure wave created by the bow of a boat or the head of a large whale.
Breaching	See page 128.
Carapace	The shell of a tortoise or turtle.
Cere	The fleshy area at the base of the bill.
Cetacean	Generic term for the Order Cetacea which includes all of the whales, dolphins and porpoises.
Crepuscular	Active in the dim light of dusk and dawn.
Eclipse plumage	The dull plumage which briefly replaces the full adult breeding plumage each year; in the case of ducks, males acquire a female-type plumage.
Endemic	Restricted to a particular geographical region.
Falcate	Hooked or curved like a sickle.
Fluking	See page 128.
Frontal shield	Area of often brightly coloured bare skin on the forehead of a bird.
Gonydeal angle	The angle of the ridge formed by the junction of the two halves of a bird's lower mandible, near the tip of the bill.
Gular	Pertaining to the throat.
Hirundine	A member of the swallow and martin family.
Inshore	Close to shore.
Kleptoparasite	An individual that forcibly steals food items obtained by another individual, often through violent harassment for prolonged periods.
Lobtailing	Common behaviour in cetaceans when the tail fluke is lifted out of the water and slapped powerfully down on the surface.
Logging	See page 129.
Melon	Bulbous-shaped forehead on a cetacean.
Migrant	A species that undertakes periodic movements to and from a given area, usually along well-defined routes and at predictable times of the year.
Offshore	Some distance from shore.
Passerine	A member of the Order Passeriformes – generally small perching birds with the large first toe pointing backwards and the outer three toes forward.
Pelagic	Term used to descibe a species that spends all or most of its life at sea.
Plastron	The hard plate which forms the underside of a tortoise or turtle.
Pod	A co-ordinated group of cetaceans. Generally used only in reference to the toothed cetaceans.
Porpoising	See page 129.
Resident	Remaining throughout the year in a given area.
Spy-hopping	See page 128.
Subspecies	A population of a species which differs in size and/or appearance from other populations of the same species (also known as a 'race').
Vagrant	An individual that wanders outside the normal range of the species.

FURTHER READING

BOWMAN, R.I., BERSON, M. AND LEVITON, A.E. (EDS) 1983. *Patterns of Evolution in Galápagos Organisms*. American Association for the Advancement of Science, Pacific Division, San Francisco, USA.

BURGHARDT, G.M. AND RAND, A.S. (EDS) 1983. *Iguanas of the World: Their Behavior, Ecology, and Conservation*. Noyes Publications, New Jersey, USA.

CARWARDINE, M. 1995. *Whales, Dolphins and Porpoises: the visual guide to all the world's cetaceans*. Dorling Kindersley, London, UK.

CASTRO, I. AND PHILLIPS, A. 1996. *A Guide to the Birds of the Galápagos Islands*. Christopher Helm/ A & C Black, London, UK.

CEPEDA, F. AND CRUZ, J.B. 1994. Status and management of seabirds on the Galápagos Islands, Ecuador. Pp 268-278 in D.N. Nettleship, J. Burger and M. Gochfeld (eds) *Seabirds on islands: threats, case studies and action plans*. BirdLife Conservation Series No. 1. BirdLife International, Cambridge, UK.

CLARK, D.B. 1980. Population ecology of an endemic neotropical island rodent *Oryzomys bauri* of Santa Fé Island, Galápagos. *J. Animal Ecology* 49:185–198.

COLLAR, N.J., GONZAGA, L.P., KRABBE, N., MADROÑO NIETO, A., NARANJO, L.G., PARKER T.A. AND WEGE, D.C. 1992. *Threatened Birds of the Americas: the ICBP/IUCN Red Data Book*. International Council for Bird Preservation, Cambridge, UK.

CONSTANT, P. 1995. *The Galápagos Islands*. Odyssey Illustrated Guide. The Guidebook Company, Hong Kong.

DE ROY, T. 1998. *Galápagos Islands Born of Fire*. Airlife, England.

GRANT, P.R. 1986 (1999). *Ecology and Evolution of Darwin's Finches*. Princeton University Press, New Jersey, USA.

HARRIS, M. 1974 (1982). *A Field Guide to the Birds of Galápagos*. Collins, London, UK.

HARRISON, P. 1983 (1993). *Seabirds: an identification guide*. Croom Helm, Beckenham, UK.

HORWELL, D. AND OXFORD, P. 1999. *Galápagos Wildlife: a visitor's guide*. Bradt Publications, UK.

JACKSON, M.H. 1993. *Galápagos, a natural history*. University of Calgary Press, Canada.

LACK, D. 1947. *Darwin's Finches*. Cambridge University Press, UK.

LEATHERWOOD, S., EVANS, W.E. AND RICE, D.W. 1972. *The Whales, Dolphins and Porpoises of the Eastern North Pacific: A Guide to their Identification in the Water*. Naval Undersea Research and Development Center, San Diego.

MARCHANT, J., PRATER, T. AND HAYMAN, P. 1986. *Shorebirds: an identification guide to the waders of the world*. Croom Helm, Beckenham, UK.

PERRY, R. (ED) 1984. *Key Environments: Galapagos*. Pergamon Press, Oxford, UK.

PRITCHARD, P.C.H. 1996. *The Galapagos Tortoises: Nomenclature and Survival Status*. Chelonian Research Monographs No. 1. Chelonian Research Foundation, Lunenberg, Massachusetts, USA.

STEADMAN, D.W. AND ZOUSMER, S. 1988. *Galápagos: Discovery on Darwin's Islands*. Airlife, England.

WEINER, J. 1994. *The Beak of the Finch: A History of Evolution in Our Time*. Alfred A. Knopf, New York, USA.

ELECTRONIC MEDIA

CHARLES DARWIN FOUNDATION INC.
http://gorp.com/igtoa/darwin.htm

CHARLES DARWIN RESEARCH STATION (CDRS)
http://www.darwinfoundation.org

FUNDACIÓN GALÁPAGOS
http://www.ecuadorable.com/galapagos

GALÁPAGOS CONSERVATION TRUST
http://www.gct.org

VIRTUAL GALÁPAGOS
http://www.terraquest.com/galapagos

GALÁPAGOS - THE CD ROM
http://naturalist.net/cdrom/

CONTACT ADDRESSES

CHARLES DARWIN FOUNDATION INC.
100 N Washington Street, Suite 311, Falls Church, VA 22046, USA.
Tel: (703) 538 6833; Fax: (703) 538 6835
Email: darwin@galapagos.org

CHARLES DARWIN RESEARCH STATION (CDRS)
Puerto Ayora, Isla Santa Cruz, Galápagos.
Tel: (703) 538 6833; Fax: (703) 538 6835
Email: cdrs@fcdarwin.org.ec

GALÁPAGOS CONSERVATION TRUST
5 Derby Street, london W1Y 7HD
Tel: (020) 7629 5049; Fax: (020) 7629 4149
Email: gct@gct.org

PHOTOGRAPHIC AND ARTWORK CREDITS

Each of the photographs and illustrations used in this book is listed in this section. As much information as possible is given, including, where known, details of the location and date, as well as the name of the photographer or artist.

The habitats of the Galápagos

Page 12. **Seabird feeding frenzy:** off Roca Redonda; August 1995; Andy Swash.

Page 12. **Rocky islet:** Roca Redonda; August 1995; Andy Swash.

Page 13. **Seabird colony:** Genovesa; January 1996; Andy Swash.

Page 13. **Rocky shore:** Fernandina; August 1995; Andy Swash.

Page 13. **Sandy beach and mangroves:** Black Beach, Isabela; August 1995; Andy Swash.

Page 14. **Mangroves:** Turtle Cove, Santa Cruz; August 1995; Andy Swash.

Page 14. **Lagoon bordered by Saltbush:** Rábida; January 1996; Andy Swash.

Page 14. **Prickly Pear *Opuntia* cacti and Sea Purslane *Sesuvium*:** South Plaza; August 1995; Andy Swash.

Page 15. **Palo Santo zone, dry season:** Darwin's Lagoon, Isabela; August 1995; Andy Swash.

Page 15. **Transition zone, agricultural area:** Santa Cruz; August 1995; Andy Swash.

Page 15. **Transition zone, disused agricultural area:** Floreana; August 1995; Andy Swash.

Page 16. ***Scalezia* zone:** Santa Cruz; January 1996; Andy Swash.

Page 16. **Fern-sedge zone:** Isabela; January 1996; Andy Swash.

The birds of the Galápagos

Page 18. **Swallow-tailed Gull:** Adult, in flight; South Plaza, Galápagos; August 1995; Andy Swash.

Page 18. **Semipalmated Plover:** Adult; Jamaica Bay, New York, USA; August; Arthur Morris (Windrush).

Page 18. **Large Ground Finch:** Adult female; Genovesa, Galápagos; January 1996; Andy Swash.

Page 18. **Chatham Mockingbird:** Adult, on branch; San Cristóbal, Galápagos; August 1995; Andy Swash.

Page 20. **Galápagos Penguin:** Adult; Bartolomé, Galápagos; August 1995; Andy Swash.

Page 21. **Flightless (or Galápagos) Cormorant:** Adult; Isabela, Galápagos; August 1995; Andy Swash.

Page 21. **Brown Pelican:** Adult; Fernandina, Galápagos; August 1995; Andy Swash.

Page 21. **Waved Albatross:** Adult; Española, Galápagos; August 1995; Andy Swash.

Page 21. **Audubon's Shearwater:** Adult; at sea off Roca Redonda, Galápagos; August 1995; Andy Swash.

Page 22. **Dark-rumped Petrel:** Adult; at sea off Isabela, Galápagos; August 1995; Andy Swash.

Page 22. **Elliot's Storm-petrel:** Adult; off Roca Redonda, Galápagos; August 1995; Andy Swash.

Page 22. **Magnificent Frigatebird:** Adult male; North Seymour, Galápagos; August 1995; Andy Swash.

Page 22. **Nazca Booby:** Adult; Genovesa, Galápagos; August 1995; Andy Swash.

Page 23. **South Polar Skua:** Adult, in flight; Half Moon Island, Antarctica; January 1993; John Cooper.

Page 23. **Swallow-tailed Gull:** Adult and chick; Genovesa, Galápagos; January 1996; Andy Swash.

Page 23. **Common Noddy:** Adult; Michaelmas Cay, Great Barrier Reef, Queensland, Australia; September 1989; Andy Swash.

Page 23. **Red-billed Tropicbird:** Adult; South Plaza, Galápagos; August 1995; Andy Swash.

Page 24. **Pied-billed Grebe:** Adult; Sanibel Island, Florida, USA; April 1989; Gordon Langsbury.

Page 24. **White-cheeked Pintail:** Adults; Santa Cruz, Galápagos; August 1995; Andy Swash.

Page 24. **Galápagos Rail:** Adult; Isabela, Galápagos; August 1995; Andy Swash.

Page 25. **Greater Flamingo:** Adults; Floreana, Galápagos; May 1997; Bruce Hallett.

Page 25. **Lava (or Galápagos) Heron:** Adult; Santiago, Galápagos; August 1995; Andy Swash.

Page 25. **American Oystercatcher:** Adult; Fernandina, Galápagos; August 1995; Andy Swash.

Page 26. **Black-necked Stilt:** Adults; Chomes, Costa Rica; February 1994; Andy Swash.

Page 26. **Semipalmated Plover:** Adult, non-breeding; North Seymour, Galápagos; August 1995; Andy Swash.

Page 26. **Black-bellied (or Grey) Plover:** Juvenile; Zach's Bay, New York, USA; October 1986; Arthur Morris (Windrush).

Page 26. **Whimbrel:** Adult; Chomes, Costa Rica; February 1991; Andy Swash.

Page 26. **Wandering Tattler:** Adult; Isabela, Galápagos; August 1995; Andy Swash.

Page 26. **Red-necked Phalarope:** Juvenile; Cornwall, UK; September 1993; Andy Swash.

Page 27. **Osprey:** Juvenile; Fort Myers Beach, Florida, USA; May 1998; Gordon Langsbury.

Page 27. **Galápagos Hawk:** Immature; Santiago, Galápagos; August 1995; Andy Swash.

Page 27. **Peregrine:** Adult; West Sussex, UK; June 1998; Gordon Langsbury.

Page 28. **Barn Owl:** Adult; illustration by Ian Lewington.

Page 28. **Short-eared Owl:** Adult; Santa Cruz, Galápagos; August 1995; Andy Swash.

Page 28. **Common Nighthawk:** Adult male; Texas, USA; April 1989; Andy Swash.

Page 29. **Galápagos Dove:** Adult; Genovesa, Galápagos; January 1996; Andy Swash.

Page 29. **Belted Kingfisher:** Adult female; Jamaica Bay, New York, USA; August; Arthur Morris (Windrush).

Page 29. **Dark-billed Cuckoo:** Adult; Palomitas, Argentina; February 1995; John Cooper.

Page 30. **Chimney Swift:** Illustration by Ian Lewington.

Page 30. **Galápagos Martin:** Adult male; illustration by Ian Lewington.

Page 31. **Vermilion Flycatcher:** Adult male; Santa Cruz, Galápagos; August 1995; Andy Swash.

Page 31. **Yellow Warbler:** Adult male; Santa Fé, Galápagos; August 1995; Andy Swash.

Page 31. **Red-eyed Vireo:** Adult; Point Pelee, Ontario, Canada; May; Arthur Morris (Windrush).

Page 31. **Bananaquit:** Adult; Trinidad; April 1998; Yvonne Dean.

Page 31. **Cedar Waxwing:** Adult; Riding Mountain NP, Manitoba, Canada; June 1997; Brayton Holt.

Page 31. **Rose-breasted Grosbeak:** Adult female; Ontario, Canada; May; Arthur Morris (Windrush).

Page 32. **Summer Tanager:** Adult female; La Selva, Costa Rica; February 1991; Andy Swash.

Page 32. **Bobolink:** Juvenile; Devon, UK; September 1991; David Tipling (Windrush).

The bird plates

All the photographs used in the plates are listed in this section; the species are listed in alphabetical order:

Blue-footed Booby: Adult, in flight; Baltra Harbour, Galápagos; August 1997; Phil Hansbro.
Blue-footed Booby: Juvenile; San Cristóbal, Galápagos; August 1995; Andy Swash.
Brown Booby: Adult, in flight; San Salvador, Bahamas; May 1995; Bruce Hallett.
Nazca Booby: Adult; Genovesa, Galápagos; January 1996; Andy Swash.
Nazca Booby: Adult, in flight showing underside; Daphne Major, Galápagos; January 1996; Andy Swash.
Nazca Booby: Adult, in flight showing upperside; Española, Galápagos; May 1997; Bruce Hallett.
Nazca Booby: Juvenile; Genovesa, Galápagos; August 1995; Andy Swash.
Red-footed Booby: Adult, brown form; Genovesa, Galápagos; August 1995; Andy Swash.
Red-footed Booby: Adult, white form; Genovesa, Galápagos; August 1995; Andy Swash.
Red-footed Booby: Adult, brown form, in flight; Genovesa, Galápagos; November 1994; Ralph Todd.
Red-footed Booby: Adult, white form, in flight; Kevin Carlson (Windrush).
Red-footed Booby: Juvenile; Genovesa, Galápagos; August 1995; Andy Swash.

Plate 9
Lava Gull: Adult; North Seymour, Galápagos; August 1995; Andy Swash.
Lava Gull: First-winter, in flight; Galápagos; December 1987; Alan Tate.
Lava Gull: Juvenile; Isabela, Galápagos; August 1995; Andy Swash.
Pomarine Jaeger/Skua: Adult, non-breeding, pale phase, in flight; off Senegal; October 1995; Dick Newell.
Pomarine Jaeger/Skua: Adult, breeding, pale phase, in flight; Taymyr, Russia; July 1994; Brayton Holt.
Pomarine Jaeger/Skua: Juvenile; Fleetwood, Lancashire, UK; April 1995; Steve Young.
Pomarine Jaeger/Skua: Juvenile, in flight; off St Simon's Island, Georgia, USA; October 1997; Giff Beaton.
South Polar Skua: Adult, in flight; Half Moon Island, Antarctica; January 1993; John Cooper.
Swallow-tailed Gull: Adult; Genovesa, Galápagos; January 1996; Andy Swash.
Swallow-tailed Gull: Adult, in flight; South Plaza, Galápagos; August 1995; Andy Swash.
Swallow-tailed Gull: Juvenile; Genovesa, Galápagos; August 1995; Andy Swash.

Plate 10
Franklin's Gull: Adult, breeding, in flight; Benton Lakes, Montana, USA; June; Arthur Morris (Windrush).
Franklin's Gull: Adult, non-breeding; Namibia; February 1999; Alan Tate.
Franklin's Gull: First-winter; Tempisque River, Costa Rica; February 1996; Andy Swash.
Kelp Gull: Adult, breeding; Lambert's Bay, South Africa; October 1997; Andy Swash.
Kelp Gull: Adult, breeding, in flight; Dunedin; October 1996; Phil Hansbro.
Kelp Gull: First-winter; Chile; February 1990; Alan Tate.
Kelp Gull: First-winter, in flight; Chatham Island; October 1996; Phil Hansbro.
Kelp Gull: Second-summer; Chile, November 1992; Roger Charlwood.
Laughing Gull: Adult, breeding; Florida, USA; March 1987; Andy Swash.
Laughing Gull: Adult, breeding, in flight; Galveston Texas, USA; April 1993; Gordon Langsbury.
Laughing Gull: Adult, non-breeding; Sanibel Island, Florida, USA; December 1994; Gordon Langsbury.

Laughing Gull: First-winter; Tempisque River, Costa Rica; February 1996; Andy Swash.
Laughing Gull: First-winter, in flight; Tempisque River, Costa Rica; February 1996; Andy Swash.

Plate 11
Common Noddy: Adult; Michaelmas Cay, Great Barrier Reef, Queensland, Australia; October 1989; Andy Swash.
Common Noddy: Adult, in flight; Dry Tortugas, Florida, USA; April 1991; Gordon Langsbury.
Red-billed Tropicbird: Adult, in flight; South Plaza, Galápagos; August 1995; Andy Swash.
Sooty Tern: Adult; Michaelmas Cay, Great Barrier Reef, Queensland, Australia; October 1989; Andy Swash.
Sooty Tern: Adult, in flight; Michaelmas Cay, Great Barrier Reef, Queensland, Australia; October 1989; Andy Swash.
Sooty Tern: Juvenile; Michaelmas Cay, Great Barrier Reef, Queensland, Australia; October 1989; Andy Swash.

Plate 12
Black Tern: Adult, breeding; Seaforth, Merseyside, UK; May 1997; Steve Young.
Black Tern: Adult, non-breeding, moulting, in flight; Crosby, Merseyside, UK; September 1994; Steve Young.
Black Tern: First-winter; Gambia; February; Barry Hughes (Windrush).
Common Tern: Adult, breeding, in flight; Fort Myers Beach, Florida, USA; April 1989; Gordon Langsbury.
Common Tern: Adult, breeding; Great Gull Island, New York, USA; June; Arthur Morris (Windrush).
Common Tern: First-winter; Kommetjie, South Africa; October 1997; Andy Swash.
Royal Tern: Adult, non-breeding; Florida, USA; January 1990; Paul Doherty.
Royal Tern: Adult, non-breeding, in flight; Florida, USA; October 1988; Paul Doherty.
White Tern: Adult, in flight; South Pacific; March; Alan Tate.

Plate 13
American Coot: Adult; Brazos Bend, Texas, USA; February 1995; Andy Swash
Black-bellied Whistling Duck: Texas, USA; A & E Morris (Windrush).
Blue-winged Teal: Adult male and female, breeding; Florida, USA; March; David Tipling (Windrush).
Masked Duck: Adult female; Brazos Bend, Texas, USA; February 1995; Andy Swash.
Masked Duck: Adult male; Brazos Bend, Texas, USA; February 1995; Andy Swash.
Pied-billed Grebe: Adult; Sanibel Island, Florida, USA; April 1989; Gordon Langsbury.
White-cheeked (or Galápagos) Pintail: Adult; Santa Cruz, Galápagos; August 1995; Andy Swash.

Plate 14
American Purple Gallinule: Adult; Brazos Bend, Texas, USA; April 1993; Gordon Langsbury.
American Purple Gallinule: Juvenile; Loxahatchee NWR, Florida, USA; February 1995; Giff Beaton.
Common Gallinule (or Moorhen): Adult; Florida, USA; April 1988; Gordon Langsbury.
Common Gallinule (or Moorhen): Juvenile; Lake Woodruff NWR, Florida USA; Arthur Morris (Windrush).
Galápagos Rail: Adult; Isabela, Galápagos; August 1995; Philip Newbold.
Paint-billed Crake: Adult; illustration by Ian Lewington.
Sora (Rail): Adult, breeding; Merritt Island, Florida, USA; April 1998; Gordon Langsbury.
Sora (Rail): Immature; Texas, USA; January 1995; Andy Swash.

Plate 15

Great Blue Heron: Adult; Lake Okeechobee, Florida, USA; March 1990; Rob Still.

Great Blue Heron: Juvenile; Santa Cruz, Galápagos; August 1995; Andy Swash.

Greater Flamingo: Adult; Isabela, Galápagos; August 1995; Andy Swash.

Greater Flamingo: Adults in flight; Eilat, Israel; July 1988; Hadoram Shirihai. Digitally manipulated.

Greater Flamingo: Immature, feeding; Coto Doñana, Spain; April 1992; Andy Swash.

Little Blue Heron: Adult; Florida, USA; April 1987; Andy Swash.

Tricolored Heron: Adult; Florida, USA; February 1994; Andy Swash.

Plate 16

Cattle Egret: Adult, breeding; Florida, USA; April 1987; Andy Swash.

Cattle Egret: Adult, non-breeding; Israel; December 1997; Dave Nye.

Great Egret: Adult; The Gambia; January 1991; Gordon Langsbury.

Little Blue Heron: Juvenile; Florida, USA; February 1994; Andy Swash.

Snowy Egret: Adult, breeding; Florida, USA; April 1987; Andy Swash.

Plate 17

Black-crowned Night-heron: Adult; The Gambia; John Geeson.

Black-crowned Night-heron: Juvenile; Costanera Sur, Argentina ; January 1993; Andy Swash.

Lava (or Galápagos) Heron: Adult; Fernandina, Galápagos; August 1995; Andy Swash.

Lava (or Galápagos) Heron: Juvenile; Fernandina, Galápagos; August 1995; Andy Swash.

Striated Heron: Adult; Hato Piñero, Venezuela; February 1992; Andy Swash.

Striated Heron: Juvenile; South Sanai, Israel; February 1987; Hadoram Shirihai.

Yellow-crowned Night-heron: Adult; Genovesa, Galápagos; January 1996; Andy Swash.

Yellow-crowned Night-heron: Immature; Genovesa, Galápagos; May 1997; Bruce Hallett.

Yellow-crowned Night-heron: Juvenile; Arthur Morris (Windrush).

Plate 18

American Oystercatcher: Adult; Fernandina, Galápagos; August 1995; Andy Swash.

American Oystercatcher: Adult, in flight; Florida, USA; September 1988; Paul Doherty.

Black Turnstone: Adult, breeding; California, USA; April 1987; Andy Swash.

Black-necked Stilt: Adult; Chomes, Costa Rica; February 1991; Andy Swash.

Black-necked Stilt: Adult, in flight; Salton Sea, California, USA; Arthur Morris (Windrush).

Ruddy Turnstone: Adult male, breeding; Sanibel Island, Florida, USA; April 1998; Gordon Langsbury.

Ruddy Turnstone: Juvenile; The Netherlands; October 1997; Gordon Langsbury.

Ruddy Turnstone: Juvenile moulting into first-winter, in flight; Kent, UK; November 1990; Paul Doherty.

Surfbird: Adult, breeding; California, USA; April 1987; Andy Swash.

Plate 19

American Golden Plover: Adult, breeding; Churchill, Manitoba, Canada; June; Arthur Morris (Windrush).

American Golden Plover: Juvenile; UK; October; David Tipling (Windrush).

Black-bellied (or Grey) Plover: Juvenile; Zach's Bay, New York, USA; October 1986; Arthur Morris (Windrush).

Black-bellied (or Grey) Plover: Adult, breeding; Richard Revels (Windrush).

Killdeer: Adult, non-breeding; Sanibel Island, Florida, USA; December 1994; Gordon Langsbury.

Pacific Golden Plover: Juvenile, moulting; Hawaii; December; A & E Morris (Windrush).

Semipalmated Plover: Adult; Jamaica Bay, New York, USA; August; Arthur Morris (Windrush).

Semipalmated Plover: Adult, non-breeding; Fort Myers Beach, Florida, USA; December 1994; Gordon Langsbury.

Wilson's Plover: Adult female, breeding; Fort Myers Beach, Florida, USA; Arthur Morris (Windrush).

Wilson's Plover: Adult male, breeding; Fort Myers Beach, Florida, USA; April 1988; Gordon Langsbury.

Plate 20

Hudsonian Godwit: Adult, breeding; Churchill, Manitoba, USA; June; Arthur Morris (Windrush).

Hudsonian Godwit: Adult, non-breeding; USA; Arthur Morris (Windrush).

Marbled Godwit: Adult; San Diego, California, USA; February 1988; Arthur Morris (Windrush).

Marbled Godwit: Juvenile, in flight; Florida, USA; October 1988; Paul Doherty.

Short-billed Dowitcher: Adult, breeding; Churchill, Manitoba, Canada; June 1997; Brayton Holt.

Short-billed Dowitcher: Adult, moulting; Fort Myers Beach, Florida, USA; April 1991; Gordon Langsbury.

Short-billed Dowitcher: Adult, non-breeding, in flight; Florida, USA; January 1990; Paul Doherty.

Whimbrel: Adult; Chomes, Costa Rica; February 1991; Andy Swash.

Whimbrel: Adult, in flight; Florida, USA; September 1988; Paul Doherty.

Willet: Adult, breeding; Texas, USA; April 1989; Andy Swash.

Willet: Adult, non-breeding; Texas, USA; January 1995; Andy Swash.

Willet: Adult, in flight; Florida, USA; April 1987; Andy Swash.

Plate 21

Buff-breasted Sandpiper: Juvenile; Tacumshin, County Wexford, Ireland; August 1994; Tom Ennis (Windrush).

Greater Yellowlegs: Adult, breeding; Fort Myers Beach, Florida, USA; April 1995; Gordon Langsbury.

Greater Yellowlegs: First-winter; Bosque del Apache, New Mexico, USA; January 1999; Gordon Langsbury.

Lesser Yellowlegs: Adult summer; Merritt Island, Florida, USA; Arthur Morris (Windrush).

Pectoral Sandpiper: Adult female, breeding; Norfolk, UK; June; Tim Loseby.

Stilt Sandpiper: Adult summer; Cambridge Bay, Canada; July 1997; Chris Schenk (Windrush).

Stilt Sandpiper: Juvenile; Jamaica Bay, New York, USA; August; Arthur Morris (Windrush).

Plate 22

Red Knot: Adult, breeding; Fort Myers Beach, Florida, USA; May 1989; Gordon Langsbury.

Red Knot: Adult, non-breeding; Fort Myers Beach, Florida, USA; April 1991; Gordon Langsbury.

Sanderling: Adult summer; Cook's Beach, Jew Jersey, USA; A & E Morris (Windrush).

Sanderling: Adult, non-breeding; Isles of Scilly, UK; October 1993; Andy Swash.

Solitary Sandpiper: Adult, breeding; Merritt Island,

Florida, USA; April 1998; Gordon Langsbury.
Spotted Sandpiper: Adult, breeding; Merritt Island, Florida, USA; April 1998; Gordon Langsbury.
Spotted Sandpiper: Juvenile; Sanibel Island, Florida, USA; December 1994; Gordon Langsbury.
Wandering Tattler: Adult, breeding; Isabela, Galápagos; August 1995; Andy Swash.
Wandering Tattler: Juvenile; Green Island, Queensland, Australia; October 1989; Andy Swash.

Plate 23
Baird's Sandpiper: Adult, non-breeding; Ushuaia, Argentina; January 1993; Brayton Holt.
Baird's Sandpiper: Juvenile; Jamaica Bay, New York, USA; August; Arthur Morris (Windrush).
Least Sandpiper: Adult, breeding; Sanibel Island, Florida, USA; April 1988; Gordon Langsbury.
Least Sandpiper: Worn adult; Jamaica Bay, New York, USA; August; Arthur Morris (Windrush).
Semipalmated Sandpiper: Adult, non-breeding; Fort Myers Beach, Florida, USA; April 1989; Gordon Langsbury.
Semipalmated Sandpiper: Juvenile; Jamaica Bay, New York, USA; August; Arthur Morris (Windrush).
Western Sandpiper: Adult, non-breeding; Fort Myers Beach, Florida, USA; April 1991; Gordon Langsbury.
Western Sandpiper: Adult, non-breeding; Florida, USA; Arthur Morris (Windrush).
White-rumped Sandpiper: Adult, non-breeding; Falkland Islands; January 1992; Gordon Langsbury.
White-rumped Sandpiper: Juvenile; Madeira; October 1993; Andy Swash.

Plate 24
Red (or Grey) Phalarope: Adult female, breeding; Cambridge Bay, Canada; July 1997; Chris Schenk (Windrush).
Red (or Grey) Phalarope: Adult male, breeding; Captive, Pensthorpe Wildlife Park, UK; May 1994; Gordon Langsbury.
Red (or Grey) Phalarope: Adult, non-breeding,; Norfolk, UK; July 1993; Tim Loseby.
Red (or Grey) Phalarope: Juvenile, moulting; Angler's Country Park, Yorkshire, UK; September 1992; Paul Doherty.
Red-necked Phalarope: Adult female, breeding; Churchill, Manitoba, Canada; June; Arthur Morris (Windrush).
Red-necked Phalarope: Adult male, breeding; Fetlar, Shetland Islands, UK; June 1993; Gordon Langsbury.
Red-necked Phalarope: Adult, non-breeding; Salton Sea, California, USA; September 1993; Nigel Bean.
Red-necked and Red (or Grey) Phalaropes: Adults, non-breeding, in flight; off Isabela, Galápagos; August 1995; Andy Swash.
Red-necked Phalarope: Juvenile; Jamaica Bay, New York, USA; August; Arthur Morris (Windrush).
Wilson's Phalarope: Adult female, breeding; Montana, USA; Arthur Morris (Windrush).
Wilson's Phalarope: Adult, non-breeding; Cornwall, UK; August; David Tipling (Windrush).

Plate 25
Galápagos Hawk: Adult, in flight showing upperside; Española, Galápagos; November 1994; Ralph Todd.
Galápagos Hawk: Adult, in flight showing underside; Fernandina, Galápagos; August 1997; Phil Hansbro.
Galápagos Hawk: Immature; Santa Fé, Galápagos; August 1995; Andy Swash.
Galápagos Hawk: Juvenile; Santa Fé, Galápagos; August 1995; Andy Swash.
Osprey: Adult; Sanibel Island, Florida, USA; December 1994; Gordon Langsbury.

Osprey: Juvenile, in flight; Fort Myers Beach, Florida, USA; May 1998; Gordon Langsbury.
Peregrine: Adult female; Wales; May; Howard Nicholls (Windrush).
Peregrine: Adult, in flight; West Sussex, UK; June 1998; Gordon Langsbury.

Plate 26
Barn Owl: Adult; Baltra, Galápagos; November 1998; Ralph Todd.
Barn Owl: Adult, in flight; illustration by Ian Lewington.
Common Nighthawk: Adult male; Texas, USA; April 1989; Andy Swash.
Common Nighthawk: Adult male, in flight; Texas, USA; April 1989; Andy Swash.
Short-eared Owl: Adult; Santa Cruz, Galápagos; August 1995; Andy Swash.
Short-eared Owl: Adult, in flight; illustration by Ian Lewington.

Plate 27
Belted Kingfisher: Adult female; Jamaica Bay, New York; August; Arthur Morris (Windrush).
Belted Kingfisher: Adult male (digitally manipulated image of immature female); Bosque del Apache, New Mexico, USA; January 1998; Gordon Langsbury.
Eared Dove: Adult; J.V. Gonzales, Argentina; January 1993; Andy Swash.
Feral Pigeon: Adult; UK; February; David Tipling (Windrush).
Galápagos Dove: Adult; Genovesa, Galápagos; January 1996; Andy Swash.

Plate 28
Black-billed Cuckoo: Adult, on branch; Higbee's Beach, New Jersey, USA; Arthur Morris (Windrush).
Dark-billed Cuckoo: Adult; Palomitas, Argentina; February 1995; John Cooper.
Groove-billed Ani: Adult; La Selva, Costa Rica; February 1991; Andy Swash.
Smooth-billed Ani: Adult; Sanibel Island, Florida, USA; April 1989; Gordon Langsbury.

Plate 29
Bank Swallow (or **Sand Martin**); **Barn Swallow; Chimney Swift; Cliff Swallow; Purple Martin** and **Galápagos Martin**: illustrations by Ian Lewington.

Plate 30
Eastern Kingbird: Adult; Texas, USA; April 1989; Andy Swash.
Large-billed (or Galápagos) Flycatcher: Adult; Floreana, Galápagos; August 1995; Andy Swash.
Vermilion Flycatcher: Adult female; Floreana, Galápagos; August 1995; Andy Swash.
Vermilion Flycatcher: Adult male; Isabela, Galápagos; August 1995; Andy Swash.

Plate 31
Bananaquit: Adult; Trinidad; April 1998; Yvonne Dean.
Blackpoll Warbler: Adult male; Point Pelee, Ontario, Canada; May; A & E Morris (Windrush).
Blackpoll Warbler: Juvenile; Isles of Scilly, UK; October 1990; David Cottridge (Windrush).
Red-eyed Vireo: Adult; Point Pelee, Ontario, Canada; May; Arthur Morris (Windrush).
Yellow Warbler: Juvenile male; Genovesa, Galápagos; August 1995; Andy Swash.
Yellow Warbler: Adult male; Santa Fé, Galápagos; August 1995; Andy Swash.
Yellow Warbler: Immature; Isabela, Galápagos; May 1997; Bruce Hallett.

Plate 32
Bobolink: Adult male; Canada; Kevin Carlson (Windrush).
Bobolink: Juvenile; Devon, UK; September 1991; David Tipling (Windrush).
Cedar Waxwing: Adult; Riding Mountain NP, Manitoba, Canada; June 1997; Brayton Holt.
Indigo Bunting: Adult male; Point Pelee, Ontario, Canada; May; A & E Morris (Windrush).
Indigo Bunting: First-winter; Ramsey Island, Wales, UK; October 1996; Alan Tate.
Rose-breasted Grosbeak: Adult female; Ontario, Canada; May; Arthur Morris (Windrush).
Rose-breasted Grosbeak: Adult male; Ontario, Canada; May; Arthur Morris (Windrush).
Summer Tanager: Adult female; La Selva, Costa Rica; February 1991; Andy Swash.
Summer Tanager: Adult male; Big Bend NP, Texas, USA; May; Arthur Morris (Windrush).

Plate 33
Charles (or Floreana) Mockingbird: Adult; Champion, Galápagos; Heidi Snell.
Chatham (or San Cristóbal) Mockingbird: Adult; San Cristóbal, Galápagos; August 1995; Andy Swash.
Galápagos Mockingbird: Adult; Genovesa, Galápagos; January 1996; Andy Swash.
Hood Mockingbird: Adult; Española, Galápagos; August 1995; Andy Swash.

Plate 34
Darwin's Finches bills; measured drawings by Ian Lewington with photographs overlaid digitally. Based on specimens from The Natural History Museum, Tring, UK (the locality from which each specimen was taken is shown in brackets): Large Ground Finch (Genovesa); Medium Ground Finch (Santa Cruz); Small Ground Finch (San Cristóbal); Sharp-beaked Ground Finch (Wolf); Woodpecker Finch (Santa Cruz); Warbler Finch (Santa Cruz); Large Cactus Finch (*conirostris*) (Española); Large Cactus Finch (*propinqua*) (Genovesa); Cactus Finch (Floreana); Vegetarian Finch (Pinta); Large Tree Finch (Floreana); Medium Tree Finch (Floreana); and Small Tree Finch (San Cristóbal). The Mangrove Finch is drawn from a photograph taken by Andy Swash at Caleta Black, Isabela, Galápagos in August 1995.

Plate 35
Large Ground Finch: Adult male; Genovesa, Galápagos; January 1996; Andy Swash.
Large Ground Finch: Female; Genovesa, Galápagos; August 1995; Andy Swash.
Medium Ground Finch: Adult Male; Santa Cruz, Galápagos; December 1998; Heidi Snell.
Medium Ground Finch: Female; Santa Cruz, Galápagos; August 1995; Andy Swash.
Sharp-beaked Ground Finch: Adult male; Genovesa, Galápagos; August 1995; Andy Swash.
Sharp-beaked Ground Finch: Female; Genovesa, Galápagos; August 1995; Andy Swash.
Small Ground Finch: Adult male; Santa Cruz, Galápagos; August 1995; Andy Swash.
Small Ground Finch: Female; Santa Cruz, Galápagos; August 1995; Andy Swash.

Plate 36
Cactus Finch: Female; South Plaza, Galápagos; January 1996; Andy Swash.
Cactus Finch: Adult male; South Plaza, Galápagos; August 1995; Andy Swash.
Large Cactus Finch: Adult male *conirostris*; Española, Galápagos; August 1995; Andy Swash.
Large Cactus Finch: Female *conirostris*; Española,

Galápagos; January 1996; Andy Swash.
Large Cactus Finch: Female *propinqua*; Genovesa, Galápagos; January 1996; Andy Swash.

Plate 37
Large Tree Finch: Adult male; Isabela, Galápagos; August 1995; Andy Swash.
Medium Tree Finch: Female/immature; Floreana, Galápagos; Windrush.
Small Tree Finch: Adult male; Santa Cruz, Galápagos; August 1995; Andy Swash.
Small Tree Finch: Female/immature; Floreana, Galápagos; August 1995; Andy Swash.
Vegetarian Finch: Adult; Galápagos; Michael Gore (Windrush).

Plate 38
Mangrove Finch: Adult; Caleta Black, Isabela, Galápagos; August 1995; Andy Swash.
Warbler Finch: Adult *cinerascens*; Española, Galápagos; August 1995; Andy Swash.
Warbler Finch: Adult *mentalis*; Genovesa, Galápagos; August 1995; Andy Swash.
Warbler Finch: Adult *olivacea*; Santa Cruz, Galápagos; May 1997; Bruce Hallett.
Woodpecker Finch: Adult non-breeding; Isabela, Galápagos; August 1995; Andy Swash.
Woodpecker Finch: Adult, breeding; Santa Cruz, Galápagos; August 1995; Andy Swash.

The reptiles of the Galápagos
Page 110. **Galápagos (or Giant) Tortoise:** Adult; Santa Cruz, Galápagos; August 1995; Andy Swash.
Page 111. **Black (or Pacific Green) Turtle:** Adult; Santa Cruz, Galápagos; January 1996; Andy Swash.
Page 111. **Galápagos Lava Lizard:** Adult male; South Plaza, Galápagos; January 1996; Andy Swash.
Page 111. **Marine Iguana:** Adult male; Isabela, Galápagos; January 1996; Andy Swash.
Page 112. **Galápagos Leaf-toed Gecko:** Adult; Galápagos; Source of photo unknown despite best endeavours.
Page 112. **Española Snake:** Adult; Española, Galápagos; David Horwell.
Page 112. **Yellow-bellied (or Pelagic) Sea Snake:** Adult; Costa Rica; Phil Myers/Animal Diversity Web.

The reptile plates
The species listed for each plate are shown in alphabetical order:

Plate 39
Galápagos (or Giant) Tortoise: Adult *hoodensis*; captive, Charles Darwin Research Station, Santa Cruz, Galápagos; August 1995; Andy Swash.
Galápagos (or Giant) Tortoise: Adult *porteri*; Santa Cruz, Galápagos; August 1995; Andy Swash.
Galápagos (or Giant) Tortoise carapaces; drawings by Rob Still after PRITCHARD, C. H. (1979) *Encyclopaedia of Turtles.*

Plate 40
Black (or Pacific Green) Turtle: Adult female on beach; Bartolomé, Galápagos; January 1996; Andy Swash.
Black (or Pacific Green), Hawksbill, Leatherback and Olive Ridley Turtles carapaces and heads: Illustrations by Rob Still.

Plate 41

Land Iguana: Adult; South Plaza, Galápagos; January 1996; Andy Swash.
Land Iguana: Adult; South Plaza, Galápagos; January 1996; Andy Swash.
Marine Iguana: Adult male *venustissimus*; Española, Galápagos; January 1996; Andy Swash.
Marine Iguana: Adult male *cristatus*; Fernandina, Galápagos; January 1996; Andy Swash.
Marine Iguana: Adult male *albemarlensis*; Isabela, Galápagos; January 1996; Andy Swash.
Marine Iguana: Group of females *cristatus*; Fernandina, Galápagos; August 1995; Andy Swash.
Marine Iguana: Females *hassi*; South Plaza, Galápagos; August 1995; Andy Swash.
Marine x Land Iguana hybrid: Subadult; South Plaza, Galápagos; David Horwell.
Santa Fé Land Iguana: Adult; Santa Fé, Galápagos: January 1996; Andy Swash.
Santa Fé Land Iguana: Adult; Santa Fé, Galápagos: January 1996; Andy Swash.

Plate 42

Española Lava Lizard: Female; Española, Galápagos; August 1995; Andy Swash.
Española Lava Lizard: Male; Española, Galápagos; January 1996; Andy Swash.
Floreana Lava Lizard: Male; Gardner-near-Floreana, Galápagos; July 1999; Heidi Snell.
Galápagos Lava Lizard: Female; South Plaza, Galápagos; August 1995; Andy Swash.
Galápagos Lava Lizard: Immature; South Plaza, Galápagos; August 1997; Phil Hansbro.
Galápagos Lava Lizard: Male; South Plaza, Galápagos; January 1996; Andy Swash.
Marchena Lava Lizard: Male; Marchena, Galápagos; August 1998; Heidi Snell.
Pinzón Lava Lizard: Female; Pinzon, Galápagos; June 2000; Heidi Snell.
San Cristóbal Lava Lizard: Male; Isla Lobos, Galápagos; July 1987; Howard Snell.

Plate 43

Galápagos Leaf-toed Gecko: Adult; Isla Tortuga, Galápagos; 1999; Heidi Snell.
Galápagos Leaf-toed Gecko: Adult on lava; Galápagos; Source of photo unknown despite best endeavours.
Galápagos Leaf-toed Gecko: Adult on stones; Galápagos; David Horwell.
Gecko *Lepidodactylus lugubris*: Adult; captive; Boris Klusmeyer/Global Gecko Association.

Plate 44

Española Snake: Adult; Española, Galápagos; David Horwell.
Fernandina Snake: Adult; Fernandina, Galápagos; August 1995; Andy Swash.
Steindachner's Snake: Adult; Santa Cruz, Galápagos; David Horwell.
Yellow-bellied (or Pelagic) Sea Snake: Adult; Costa Rica; Phil Myers/Animal Diversity Web.

The mammals of the Galápagos

Page 126. **Galápagos Fur Seal:** Female; Santiago, Galápagos; August 1995; Andy Swash.
Page 126. **Large Fernandina Rice Rat:** Fernandina, Galápagos; Robert C. Dowler.
Page 127. **Galápagos Red Bat:** San Cristóbal, Galápagos; Gary F. McCracken.

Page 128. **Blue Whale:** Blow; Snaefellsness, Iceland; June 1998; Dylan Walker.
Page 128. **Humpback Whale:** Breaching; Maui, Hawaii, USA; March 1996; Lori Mazzuca.
Page 128. **Northern Bottlenose Whale:** Spy-hopping; Skye, Scotland; August 1998: Dylan Walker.
Page 128. **Humpback Whale:** Fluking; Keflavik, Iceland; June 1998; Dylan Walker.
Page 129. **Blue Whale and head:** illustration by Marc Dando from *Sealife: A Guide to the Marine Environment* (ed. G. Waller), Pica Press, UK.
Page 129. **Striped Dolphin:** illustration by Marc Dando from *Sealife: A Guide to the Marine Environment* (ed. G. Waller), Pica Press, UK.
Page 129. **Short-finned Pilot Whale:** Logging; Tenerife; September 1998; Dylan Walker.
Page 129. **Dusky Dolphin** and **Common Dolphin**: Porpoising; Kaikoura, New Zealand; March 1999; Sam Taylor.
Page 130. **Blue Whale:** Snaefellsness, Iceland; June 1998; Dylan Walker.
Page 130. **Sperm Whale:** Manado, North Sulawesi, Indonesia; August 1998; Pietro Pecchioni
Page 130. **Blainville's Beaked Whale:** Hope Town, Bahamas; March 1994; Stephanie A. Norman.
Page 130. **Short-finned Pilot Whale:** Tenerife; February 1998; Dylan Walker.
Page 131. **Long-snouted Spinner Dolphin:** Maui, Hawaii, USA; February 1998; Lori Mazzuca.

The mammal plates

The species listed for each plate are shown in alphabetical order:

Plate 45

Galápagos Fur Seal: Female with pup; Santiago, Galápagos; August 1995; Andy Swash.
Galápagos Fur Seal: Bull; Galápagos; David Horwell.
Galápagos Sea Lion: Bull; Galápagos; David Horwell.
Galápagos Sea Lion: Females; Genovesa, Galápagos; August 1995; Andy Swash.
Galápagos Sea Lion: Pup; Santiago, Galápagos; August 1995; Andy Swash.

Plate 46

Large Fernandina Rice Rat: Fernandina, Galápagos; Robert C. Dowler.
Small Fernandina Rice Rat: Fernandina, Galápagos; Robert C. Dowler.
Santa Fé Rice Rat: Santa Fé, Galápagos; 1980; Howard Snell.
Santiago Rice Rat: Santiago, Galápagos; Robert C. Dowler.

Plate 47

Black Rat: Mark Lucas.
Brown Rat: Colin Carver.
Galápagos Red Bat: San Cristóbal, Galápagos; Gary F. McCracken.
Hoary Bat: Isabela, Galápagos; Gary F. McCracken.
House Mouse: Richard Revels.

Plate 48

Bryde's Whale: Back and dorsal fin; La Paz,Baja California Sur, Mexico; Dagmar C. Fertl.
Bryde's Whale: Head and back: La Paz, Baja California Sur, Mexico; Dagmar C. Fertl.
Fin Whale: Back and dorsal fin; Monterey Bay, California, USA; August 1997; Graeme Cresswell.
Fin Whale: Right jaw; Gulf of St. Lawrence, Canada; August 1998; Graeme Cresswell.

Fin Whale: Left jaw; Gulf of St. Lawrence, Canada; August 1998; Graeme Cresswell.
Fin Whale: Blowhole and blow; Bay of Fundy, Canada; August 1998; Graeme Cresswell.
Minke Whale: Back and dorsal fin; Gulf of St. Lawrence, Canada; August 1998; Graeme Cresswell.
Minke Whale: Head and back; Gulf of St. Lawrence, Canada; August 1998; Graeme Cresswell.
Sei Whale: Near Jan Mayen; July 1998; George McCallum/ Whalephoto Berlin.
Sei Whale: Vestfjord Lofoten; September 1998; George McCallum/Whalephoto Berlin.

Plate 49
Blue Whale: Initial surface; Snaefellsness, Iceland; June 1998; Dylan Walker.
Blue Whale: Head and back; Snaefellsness, Iceland; June 1998; Dylan Walker.
Blue Whale: Back and dorsal fin; Snaefellsness, Iceland; June 1998; Dylan Walker.
Humpback Whale: Pair surfacing; Keflavik, Iceland; June 1998; Graeme Cresswell.
Humpback Whale: Keflavik, Iceland; June 1998; Dylan Walker.
Humpback Whale: Maui, Hawaii, USA; February 1996; Lori Mazzuca.
Sperm Whale: Kaikoura, New Zealand; January 1999; Sam Taylor.
Sperm Whale: Kaikoura, New Zealand; April 1999; Sam Taylor.
Sperm Whale: Manado, North Sulawesi, Indonesia; June 1998; Carla Benoldi

Plate 50
Blainville's Beaked Whale: Hope Town, Bahamas; March 1994; Stephanie A. Norman.
Blainville's Beaked Whale: Hope Town, Bahamas; March 1994; Stephanie A. Norman.
Cuvier's Beaked Whale: Female and calf; Bay of Biscay, Spain; July 1998; Dylan Walker.
Cuvier's Beaked Whale: Dark-headed individual; Bay of Biscay, Spain; July 1997; Jonathan Mitchell.
Cuvier's Beaked Whale: Pale-headed individual; Bay of Biscay, Spain; August 1998; Dylan Walker.
Dwarf/Pygmy Sperm Whale: Manado, North Sulawesi, Indonesia; August 1998; Pietro Pecchioni.
Dwarf Sperm Whale: Illustration by Marc Dando from *Sealife: A Guide to the Marine Environment* (ed. G. Waller), Pica Press, UK. Digitally manipulated.
Pygmy Sperm Whale: Illustration by Marc Dando from *Sealife: A Guide to the Marine Environment* (ed. G. Waller), Pica Press, UK. Digitally manipulated.

Plate 51
False Killer Whale: Breaching; Mercury Bay, New Zealand; January 1999; Dirk Neumann.
False Killer Whale: Surfacing; Lana'i, Hawaii, USA; April 1997; Lori Mazzuca.
Killer Whale (or Orca): Family pod; Snaefellsness, Iceland; June 1998; Graeme Cresswell.
Killer Whale (or Orca): Adult male; Kaikoura, New Zealand; January 1999; Sam Taylor.
Melon-headed Whale: Lana'i, Hawaii, USA; October 1997; Lori Mazzuca.
Melon-headed Whale: Spy-hopping; Manado, North Sulawesi, Indonesia; June 1998; Carla Benoldi.
Pygmy Killer Whale: Manado, North Sulawesi, Indonesia; June 1998; Carla Benoldi.
Short-finned Pilot Whale: Tenerife; October 1998; Dylan Walker.
Short-finned Pilot Whale: Tenerife; October 1998; Dylan Walker.

Plate 52
Common Dolphin: Porpoising; Mercury Bay, New Zealand; March 1999; Dirk Neumann.
Common Dolphin: Surfacing; Bay of Biscay, Spain; August 1999; Dylan Walker.
Long-snouted Spinner Dolphin: Porpoising; Gulf of Mexico, Dagmar C. Fertl.
Pantropical Spotted Dolphin: Porpoising; Ecuador; November 1998; Stephanie A. Norman.
Pantropical Spotted Dolphin: Surfacing; Manado, North Sulawesi, Indonesia; May 1998; Carla Benoldi.
Striped Dolphin: Porpoising; Gulf of Mexico; Dagmar C. Fertl.

Plate 53
Bottle-nosed Dolphin: Porpoising; Oahu, Hawaii, USA; April 1996; Lori Mazzuca.
Bottle-nosed Dolphin: Surfacing; Tenerife; October 1998; Dylan Walker.
Fraser's Dolphin: Surfacing pair; Manado, North Sulawesi, Indonesia; June 1998; Carla Benoldi.
Risso's Dolphin: Pod logging; Manado, North Sulawesi, Indonesia; September 1998; Carla Benoldi.
Risso's Dolphin: Pair surfacing; Manado, North Sulawesi, Indonesia; September 1998; Carla Benoldi.
Rough-toothed Dolphin: Breaching; Tahiti; May 1999; Sam Taylor.
Rough-toothed Dolphin: Surfacing; 500 miles west of Honduras; November 1998; Stephanie A. Norman.

If you believe you have better quality photos of any of the species in this guide please contact the authors at the **WILD***Guides* address or email photos@wildguides.co.uk

INDEX OF ENGLISH AND SCIENTIFIC NAMES

This index includes the common English and scientific names of all the birds, mammals and reptiles mentioned in the text.

Figures in *italicised bold red text* refer to the number(s) of the plate(s) on which the species is illustrated.
Figures in *italicised bold black text* indicate other page number(s) on which there is an illustration.
Figures in **bold text** refer to the page(s) on which the main text for the species can be found.
Figures in plain text relate to other page(s) where the species is mentioned.